Systems of Linear Inequalities

Popular Lectures in Mathematics

Survey of Recent East European Mathematical Literature

A project conducted by
Izaak Wirszup,
Department of Mathematics,
the University of Chicago,
under a grant from the
National Science Foundation

A. S.
Solodovnikov

Systems of Linear Inequalities

Translated and
adapted from the
Russian edition by
Lawrence M. Glasser and
Thomas P. Branson

The
University of Chicago
Press
Chicago and
London

The University of Chicago Press, Chicago 60637
The University of Chicago Press, Ltd., London

Printed in the United States of America
84 83 82 81 80 5 4 3 2 1

Library of Congress Cataloging in Publication Data

Solodovnikov, Aleksandr Samuilovich.
Systems of linear inequalities.

(Popular lectures in mathematics)
Translation of Sistemy lineĭnykh neravenstv.
1. Inequalities (Mathematics) I. Title.
QA295.S6413 515′.26 79-16106
ISBN 0-226-76786-8

Contents

Foreword vii

1. Several Facts from Analytic Geometry 1
2. Geometric Interpretation of a System of Linear Inequalities in Two and Three Unknowns 12
3. The Convex Hull of a Point Set 17
4. Convex Polyhedral Cones 21
5. The Solution Domain of a System of Inequalities in Two Unknowns 27
6. The Solution Domain of a System in Three Unknowns 41
7. Systems of Linear Inequalities in an Arbitrary Number of Unknowns 50
8. Inconsistent Systems 55
9. Dual Convex Polyhedral Cones 65
10. The Duality Theorem of Linear Programming 72

References 81

Foreword

A first degree (or *linear*) inequality in n unknowns $x_1, x_2, \ldots, x_n$ is a relation of the form

$$a_1x_1 + a_2x_2 + \cdots + a_nx_n + k \geq 0\,,$$

where $a_1, a_2, \ldots, a_n$ and k are constant real numbers. For example, a linear inequality in two unknowns is a relation of the form

$$ax + by + c \geq 0\,,$$

where a, b, and c are constants and x and y are unknowns. The theory of systems of linear inequalities is a small but very attractive corner of mathematics. Its interest is enhanced by the beauty of its geometric content; interpreted geometrically, a system of linear inequalities in two unknowns describes a convex polygonal region in the Euclidean plane, and a system in three unknowns determines a convex polyhedral solid in Euclidean three-dimensional space. Thus the study of convex polyhedra, a branch of geometry almost as old as recorded history, is transformed into one of the principal chapters in the theory of linear inequalities. Some features of this theory are very close to the heart of the algebraist—for example, the remarkable analogy between the properties of systems of linear inequalities and the properties of systems of linear equations. These analogies were not studied in great detail until quite recently.

Until the mid 1940s, it was generally believed that the theory of linear inequalities would continue to be a subject of purely mathematical interest. The situation changed radically, however, with the advent of *linear programming*, a new area of applied mathematics with important applications in economics and technology. From a mathematical point of view, however, linear programming is but one section (although a very important one) of the theory of systems of linear inequalities.

The purpose of this booklet is to familiarize the reader with various aspects of the theory of systems of linear inequalities, including methods of solution, geometric and algebraic considerations, and the principal problems of linear programming. No knowledge beyond what is covered in high school mathematics is prerequisite for this book. Several references are given at the end of the book, so that the interested reader may familiarize himself in more detail with various topics relating to systems of linear inequalities. This list has been supplemented by the recent monograph of S. N. Chernikov.

Before proceeding any further, let us devote a few words to the history of the subject we are about to consider. Although by its very nature the theory of linear inequalities would be expected to deal with the most basic and elementary portions of mathematics, it received relatively little consideration until quite recently.

Since the closing years of the last century, occasional works have illuminated one or another property of these systems. In this connection one must mention such mathematicians as G. Minkowski (one of the greatest geometers of the late nineteenth and early twentieth centuries, particularly famous for his work on convex sets and the creation of "Minkowski geometry"), G. F. Voronoi (one of the founders of the St. Petersburg school of number theory), A. Haar (the Hungarian mathematician who became famous for his work on "integration in groups"), and H. Weyl (one of the leading mathematicians of the first half of our century, whose life and activities are described in I. M. Yaglom's recent booklet "Hermann Weyl," published by Znanie 1967[1]). A number of their ideas are represented in this booklet without reference to their authors.

Intensive development of the theory did not actually begin until the 1940s and 1950s, when the rapid growth of allied disciplines (linear, convex, and other forms of mathematical "programming" such as "game theory") made a deep and systematic study of linear inequalities necessary. At the present time a complete list of references to books and articles on linear inequalities would probably encompass hundreds of titles.

1. See also I. M. Yaglom's article "Hermann Weyl and the Concept of Symmetry" in the Russian edition of H. Weyl, *Symmetry* (Moscow: Science Publishers, 1968), pp. 5–32. (Weyl's book—without Yaglom's preface—was published in English by Princeton University Press, 1952.)

Systems of Linear Inequalities

1 Several Facts from Analytic Geometry

1.1. Operations on Points

We introduce a Cartesian (rectilinear) coordinate system into the Euclidean plane in which each point M is represented by an ordered pair of real coordinates (x, y). If the point M has the coordinates (x, y), we write

$$M = (x, y) .$$

Specification of a coordinate system allows us to define a number of operations on points in the plane, among them the *sum* of two points and the *product* of a point with a *scalar* (real number).

First of all, we must introduce some notation. If M and N are points in the plane, the line segment joining M and N will be denoted by MN, the vector with initial point M and terminal point N will be denoted $\mathbf{MN}$, and the ray with initial point M determined by N will be called "the ray MN." The length of MN or $\mathbf{MN}$ will be denoted by $\|MN\|$. The line determined by M and N will be called "the line MN."

The addition of points is defined as follows: if $M_1 = (x_1, y_1)$ and $M_2 = (x_2, y_2)$, then

$$M_1 + M_2 = (x_1 + x_2, y_1 + y_2) .$$

Thus we add points simply by adding their corresponding coordinates.

The geometric interpretation of this operation is very simple (fig. 1.1): the point $M_1 + M_2$ is the fourth vertex of the parallelogram constructed from the segments OM_1 and OM_2 (O is the origin of the coordinate system). The other three vertices of the parallelogram are M_1, O, and M_2.

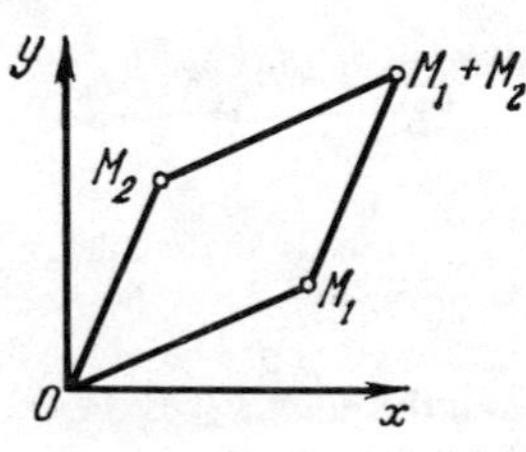

Fig. 1.1

The same idea can be expressed differently: the point $M_1 + M_2$ is obtained by the parallel displacement of the point M_2 along a segment parallel to OM_1 by an amount equal to the length OM_1 of this segment.

The multiplication of the point $M = (x, y)$ by an arbitrary real number k is defined by the rule

$$kM = (kx, ky)\,.$$

The geometric interpretation of this operation is even simpler than that of addition: for $k \geq 0$, the point $M' = kM$ lies on the ray OM in such a way that $OM' = k \cdot OM$; for $k < 0$ the point M' lies on the ray with initial point O that is directed opposite the ray OM in such a way that $OM' = |k| \cdot OM$ (fig. 1.2).

The verification of these geometric interpretations is left as a simple exercise for the reader.[1]

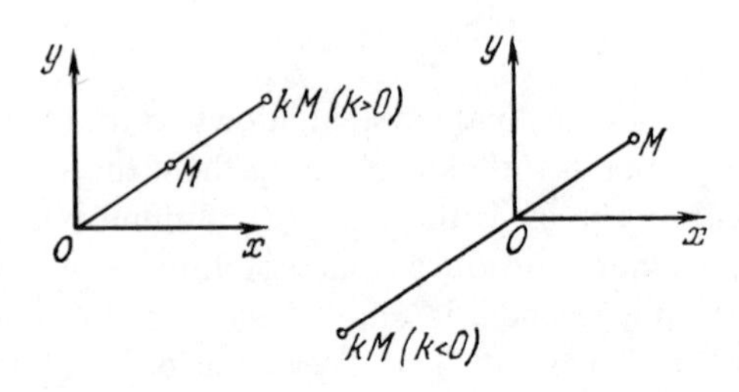

Fig. 1.2

These operations are very convenient for the purpose of translating geometric facts into algebraic language. We shall give several examples.

1. *The segment M_1M_2 consists of all points of the form*

$$s_1M_1 + s_2M_2\,,$$

where s_1 and s_2 are any two nonnegative real numbers whose sum is 1.

To prove this assertion, consider any point on the segment M_1M_2.

1. Only if the reader is not familiar with the basic ideas of vector analysis. From the point of view of vectors, our operations have the following well-known interpretation: the point $M_1 + M_2$ is the end point of the vector $OM_1 + OM_2$ and the point kM is the endpoint of the vector $k \cdot OM$ (with the provision that the initial points of these new vectors also lie at the point O).

By constructing straight lines through M parallel to the lines OM_2 and OM_1, respectively, we determine a point N_1 on the segment OM_1 and a point N_2 on the segment OM_2 (fig. 1.3). Let

$$s_1 = \frac{M_2M}{M_2M_1} \quad \text{and} \quad s_2 = \frac{M_1M}{M_1M_2};$$

the numbers s_1 and s_2 are clearly nonnegative and add up to 1. From the similarity of triangles OM_1M_2, N_2MM_2, and N_1M_1M (fig. 1.3), we find that

$$\frac{ON_1}{OM_1} = \frac{N_2M}{OM_1} = \frac{M_2M}{M_2M_1} = s_1 \quad \text{and} \quad \frac{ON_2}{OM_2} = \frac{N_1M}{OM_2} = \frac{M_1M}{M_1M_2} = s_2,$$

from which we have $N_1 = s_1M_1$ and $N_2 = s_2M_2$. But $M = N_1 + N_2 = s_1M_1 + s_2M_2$. Thus the segment is contained in the set of points $s_1M_1 + s_2M_2$. Conversely, as the point M runs across the segment M_1M_2 starting at M_1, the number s_2 takes on all values from 0 to 1 inclusive, so that $s_1 = 1 - s_2$ takes on all values from 1 to 0 inclusive. Thus the set of points $s_1M_1 + s_2M_2$ is contained in M_1M_2, so that by the above, the two sets are equal.

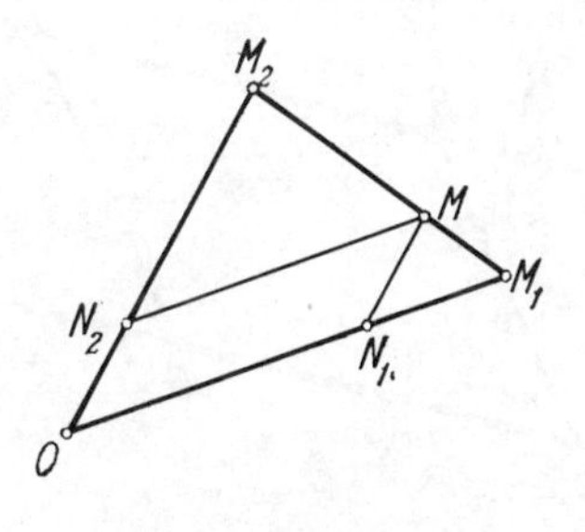

Fig. 1.3

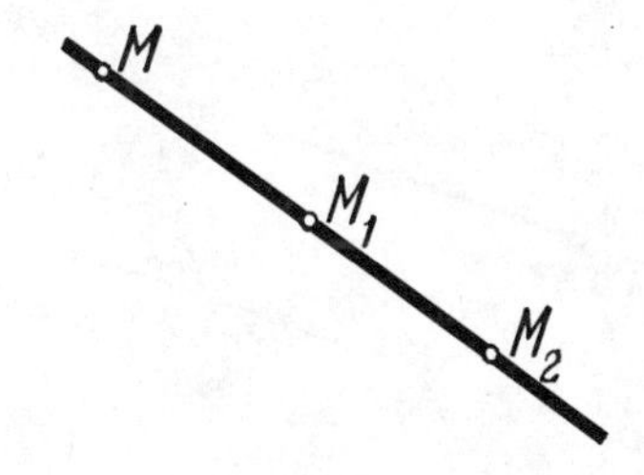

Fig. 1.4

2. *The straight line M_1M_2 is the set of points*

$$tM_1 + (1 - t)M_2,$$

where t is a real number.

Indeed, if the point M lies on the segment M_1M_2, then our assertion follows from the one just proved. If M lies outside the segment M_1M_2, then either the point M_1 lies on the segment MM_2 (as in fig. 1.4), or M_2 lies on the segment MM_1. Let us assume, for example, that the first case is realized. Then, from what we have shown,

$$M_1 = sM + (1 - s)M_2 \quad (0 < s < 1).$$

Therefore

$$M = \frac{1}{s} M_1 - \frac{1 - s}{s} M_2 = tM_1 + (1 - t)M_2,$$

where $t = 1/s$. We leave the case where M_2 lies on the segment MM_1 to the reader.

The converse inclusion is obvious by an argument similar to that above.

3. *As the parameter s increases from* 0 *to* ∞, *the point sB traces out the ray OB,*[2] *and the point $A + sB$ traces out the ray with initial point A in the direction* $\mathbf{OB}$ (fig. 1.5). *As s decreases from* 0 *to* $-\infty$, *the points sB and $A + sB$ trace out the rays supplementary* (*opposite*) *to those above* (fig. 1.6).

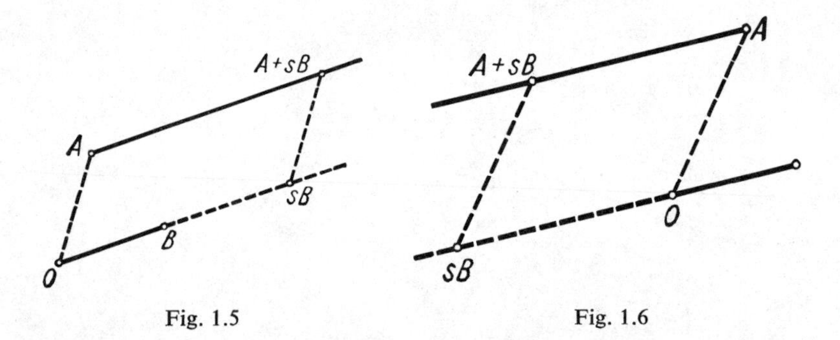

Fig. 1.5

Fig. 1.6

It follows from assertion 3 that as s increases from $-\infty$ to $+\infty$, the point $A + sB$ traces out the straight line parallel to OB which passes through A.

The operations of addition and multiplication by a scalar can, of course, be extended to points in Euclidean three-dimensional space. In

2. It is assumed that the point B is distinct from the origin O.

this case, by definition, if $M_1 = (x_1, y_1, z_1)$ and $M_2 = (x_2, y_2, z_2)$,

$$M_1 + M_2 = (x_1 + x_2, y_1 + y_2, z_1 + z_2)\,,$$
$$kM_1 = (kx_1, ky_1, kz_1)\,.$$

It is evident that all the assertions proved above remain valid for points in three-space.

To conclude this section, we shall introduce some notation and a convention which will allow us to formulate many assertions in a clear and concise way. If a point M belongs to the set $\mathscr{M}$, we write $M \in \mathscr{M}$ (M "is an element of" $\mathscr{M}$). If $\mathscr{K}$ and $\mathscr{L}$ are any two sets of points in the plane or in three-space, then by their *sum* $\mathscr{K} + \mathscr{L}$ we shall mean the set of all points $K + L$ such that $K \in \mathscr{K}$ and $L \in \mathscr{L}$.

Using the geometric interpretation of the addition of points, we can give a simple rule for the addition of two point sets $\mathscr{K}$ and $\mathscr{L}$: for each point $K \in \mathscr{K}$, we form the set obtained from $\mathscr{L}$ by parallel displacement by the vector $\boldsymbol{OK}$, and then take the union of all the sets obtained in this way. This set will be precisely $\mathscr{K} + \mathscr{L}$.

We shall give four examples.

1. Let the set $\mathscr{K}$ consist of a single point K, while $\mathscr{L}$ contains any number of points. The set $K + \mathscr{L}$ is the result of translating the set $\mathscr{L}$ by the vector $\boldsymbol{OK}$ (fig. 1.7). In particular, if $\mathscr{L}$ is a straight line, then $K + \mathscr{L}$ is a line parallel to $\mathscr{L}$. If the line $\mathscr{L}$ passes through the origin, then $K + \mathscr{L}$ is the straight line parallel to $\mathscr{L}$ passing through the point K (fig. 1.8).

2. $\mathscr{K}$ and $\mathscr{L}$ are line segments (in the plane or in three-space) which are not parallel to one another (fig. 1.9). Then the set $\mathscr{K} + \mathscr{L}$ is a

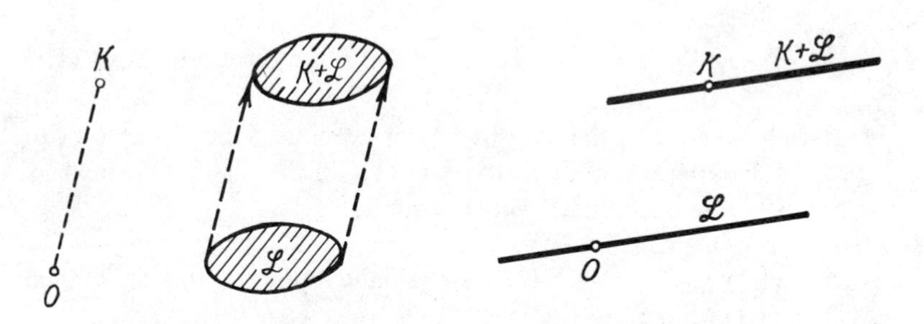

Fig. 1.7

Fig. 1.8

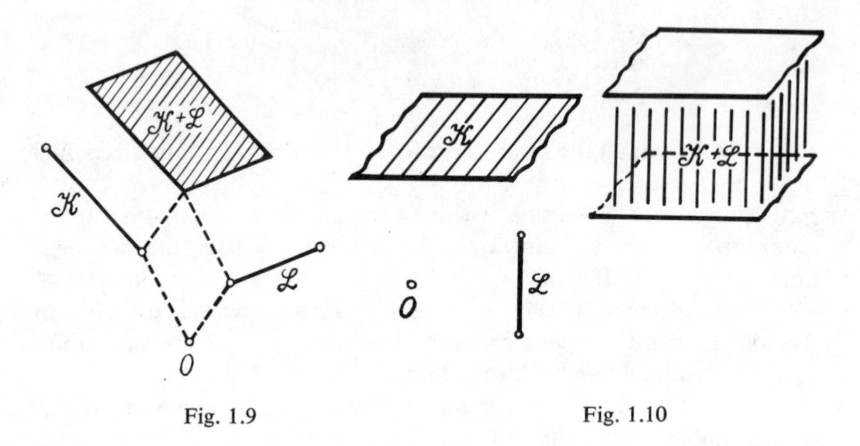

Fig. 1.9

Fig. 1.10

parallelogram whose sides are segments of length equal to, and parallel to, $\mathscr{K}$ and $\mathscr{L}$. What happens when $\mathscr{K}$ and $\mathscr{L}$ are parallel?

3. $\mathscr{K}$ is a plane (in three-space) and $\mathscr{L}$ is a line segment which is not parallel to $\mathscr{K}$. The set $\mathscr{K} + \mathscr{L}$ is the portion of space included between two planes parallel to $\mathscr{K}$ (fig. 1.10).

4. $\mathscr{K}$ and $\mathscr{L}$ are circles of radius r_1 and r_2 centered at P_1 and P_2, respectively, which lie in the same plane π. Then $\mathscr{K} + \mathscr{L}$ is the circle with radius $r_1 + r_2$ centered at the point $P_1 + P_2$ and lying in a plane parallel to π (fig. 1.11).

1.2. The Geometric Interpretation of Linear Equations and Inequalities in Two and Three Unknowns

Consider a linear equation in the two unknowns x and y:

$$ax + by + c = 0 \qquad \text{(where } a \text{ and } b \text{ are not both 0)}\,. \tag{1.1}$$

By viewing x and y as the coordinates of a point in the plane, we can pose the following question: What point set in the plane is formed by the points whose coordinates satisfy equation (1.1)?

The answer is probably known to the reader: *The point set defined by equation* (1.1) *is a straight line in the plane.* Indeed, if $b \neq 0$, then equation (1.1) reduces to the form

$$y = kx + p\,,$$

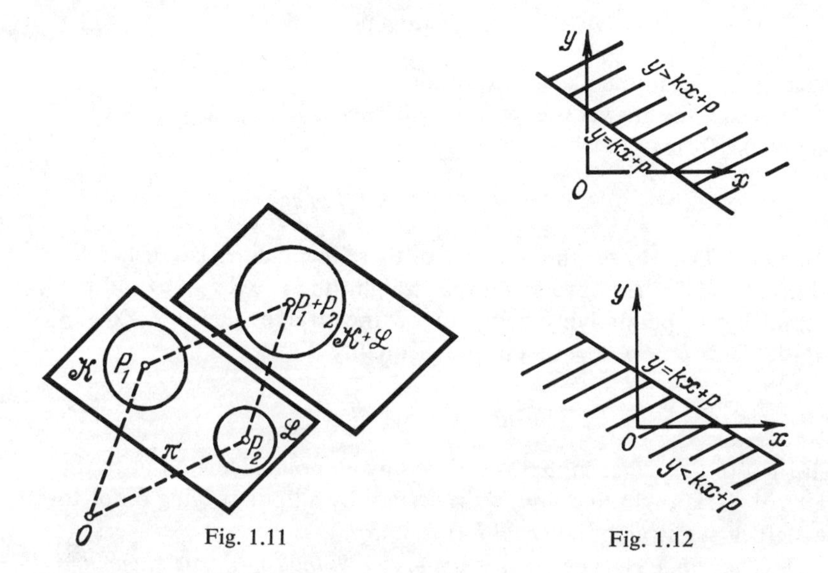

Fig. 1.11

Fig. 1.12

and it is well known that this equation determines a straight line. If, however, $b = 0$, then $a \neq 0$, and the equation reduces to

$$x = h\,,$$

and defines a straight line parallel to the y-axis.

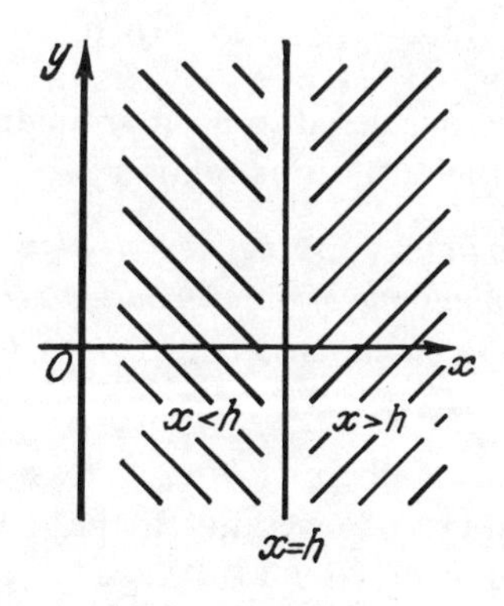

Fig. 1.13

A similar question arises in connection with the inequality

$$ax + by + c \geq 0 \qquad (a, b \text{ not both } 0)\,. \tag{1.2}$$

What point set in the plane is determined by the inequality (1.2)?

Here, too, the answer is easy. If $b \neq 0$, then the inequality reduces to one of the form

$$y \geq kx + p \quad \text{or} \quad y \leq kx + p\,.$$

It is not difficult to see that the first of these inequalities is satisfied by all points which lie above or on the straight line $y = kx + p$, and the second by all points lying below or on the straight line $y = kx + p$ (fig. 1.12). If $b = 0$, $a \neq 0$, and the inequality reduces to

$$x \geq h \quad \text{or} \quad x \leq h\,.$$

The first of these inequalities is satisfied by all points lying on or to the right of the straight line $x = h$, the second by all points lying on or to the left of the straight line $x = h$ (fig. 1.13).

Thus, *equation* (1.1) *defines a straight line in the coordinate plane, and the inequality* (1.2) *determines one of the two half-planes into which this straight line divides the plane* (*we presume that the line itself belongs to both of these half-planes*).

We now wish to answer the same questions with respect to the equation

$$ax + by + cz + d = 0 \qquad (a, b, c \text{ not all } 0)\,, \tag{1.3}$$

and the inequality

$$ax + by + cz + d \geq 0 \qquad (a, b, c \text{ not all } 0)\,, \tag{1.4}$$

where x, y, and z are considered to be coordinates in three-space. It is not hard to see that the result is as follows.

THEOREM 1.1. *Equation* (1.3) *defines a plane in three-space, and inequality* (1.4) *determines one of the two half-spaces into which this plane divides three-space* (*the plane itself is presumed to belong to both half-spaces*).

Proof. Of the three numbers a, b, c, at least one is different from zero; assume, without loss of generality, that $c \neq 0$. Then equation (1.3) reduces to one of the form

$$z = kx + my + p\,. \tag{1.5}$$

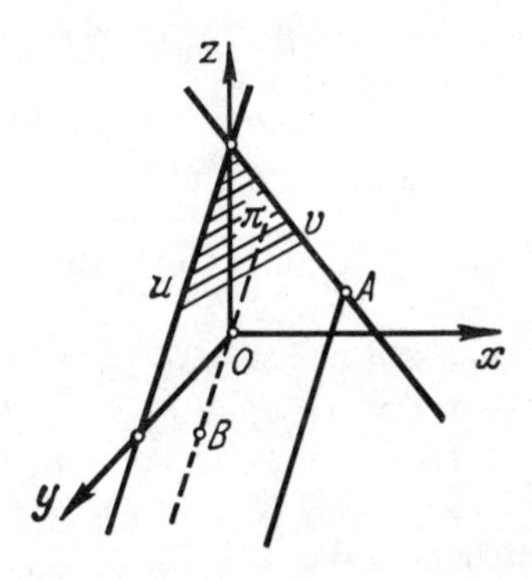

Fig. 1.14

Let $\mathscr{L}$ denote the set of all points $M = (x, y, z)$ which satisfy equation (1.5). Our purpose is to show that $\mathscr{L}$ is a plane.

Let us determine which points of $\mathscr{L}$ belong to the yz-plane (the plane $x = 0$) by setting $x = 0$ in equation (1.5). We find that

$$z = my + p\,. \tag{1.6}$$

Thus, the intersection of $\mathscr{L}$ with the yz-plane is a straight line u, defined in this plane by equation (1.6) (fig. 1.14).

In the same way we find that the intersection of $\mathscr{L}$ with the xz-plane is a straight line v, defined in this plane by the equation

$$z = kx + p\,. \tag{1.7}$$

Thus the two straight lines u and v must pass through the point

$$P = (0, 0, p)\,.$$

We shall let π denote the plane determined by the straight lines u and v, and show that π is contained in the set $\mathscr{L}$.

For this purpose, we need only establish the following fact: any straight line parallel to u which passes through any point $A \in v$ is contained in $\mathscr{L}$.

Let us first pick some point B such that the line OB is parallel to u. In the yz-plane, the equation $z = my + p$ defines the line u; this means that the equation $z = my$ defines a line parallel to u passing through the origin (shown as a dotted line in fig. 1.14). We may take as our point B the point with coordinates $y = 1$, $z = m$, which lies on this line.

Any point $A \in v$ has the coordinates $(x, 0, kx + p)$. The point B we have selected has the coordinates $(0, 1, m)$, so that the straight line parallel to u which passes through A consists of the points

$$A + sB = (x, 0, kx + p) + s(0, 1, m) = (x, s, kx + p + sm),$$

where s is any real number (see assertion 3 of sec. 1.1). It is easy to verify that the coordinates of the point $A + sB$ satisfy equation (1.5), which means that $A + sB \in \mathcal{L}$, proving that the plane π (which is the union of all straight lines parallel to u which pass through some point $A \in v$) is entirely contained in the set $\mathcal{L}$.

The remaining step is to show that $\mathcal{L}$ is *identical* to π—that is, to show that the set $\mathcal{L}$ contains no point which lies outside of π.

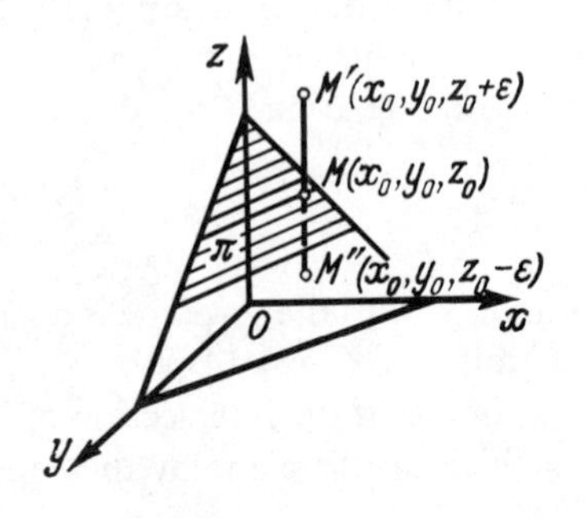

Fig. 1.15

To accomplish this we consider three points: a point $M = (x_0, y_0, z_0)$, which lies in the plane π, a point $M' = (x_0, y_0, z_0 + \varepsilon)$ lying "above" the plane π $(\varepsilon > 0)$, and a point $M'' = (x_0, y_0, z_0 - \varepsilon)$ lying "below" π (fig. 1.15). Because $M \in \pi$, $z_0 = kx_0 + my_0 + p$, and consequently

$$z_0 - \varepsilon < kx_0 + my_0 + p < z_0 + \varepsilon .$$

From this it is clear that the coordinates of the point M' satisfy the strict inequality

$$z > kx + my + p ,$$

and the coordinates of the point M'' satisfy the strict inequality

$$z < kx + my + p\,.$$

Therefore, M' and M'' do not belong to $\mathscr{L}$, and since all points not on π are of the form M' or M'' (that is, are either "above" or "below" π), $\mathscr{L}$ must coincide with π. Moreover, it follows from our considerations that the set of all points satisfying the inequality (1.4),

$$ax + by + cz + d \geq 0\,,$$

is one of the two half-spaces ("upper" or "lower") into which the plane π divides all of three-space.

2 The Geometric Interpretation of a System of Linear Inequalities in Two or Three Unknowns

Consider the system of inequalities

$$\begin{aligned} a_1x + b_1y + c_1 &\geq 0 \\ a_2x + b_2y + c_2 &\geq 0 \\ \cdots\cdots\cdots&\cdots\cdots \\ a_mx + b_my + c_m &\geq 0 \end{aligned} \tag{2.1}$$

in the two unknowns x and y.

The first inequality of the system defines a half-plane Π_1 in the coordinate plane, the second defines a half-plane Π_2, and so on. If some ordered pair (x, y) satisfies all the inequalities (2.1), then the corresponding point $M = (x, y)$ will belong to all the half-planes $\Pi_1, \Pi_2, \ldots, \Pi_m$ simultaneously. In other words, the point M belongs to the *intersection* of these half-planes. It is clear that the intersection of a finite number of half-planes is some region $\mathcal{K}$ such as the one shown in figure 2.1. Along the boundaries defining the region we have indicated the interior by shading. The arrows indicate on which side of a given line the corresponding half-plane lies.

The region $\mathcal{K}$ is called the *solution domain* of the system (2.1). Note that the solution domain is not necessarily bounded; the intersection of several half-planes can form an unbounded region as shown in figure 2.2. In view of the fact that the boundary of such a region consists of line segments and/or rays, we shall call $\mathcal{K}$ a *polygonal region.* (When $\mathcal{K}$ is bounded, we call it simply the *solution polygon*[1] of the system (2.1).) Of course, it can also happen that there is no point which belongs simultaneously to all the half-planes, that is, the solution domain is

1. Such solution polygons are the union of a polygon (in the usual sense) with its interior.

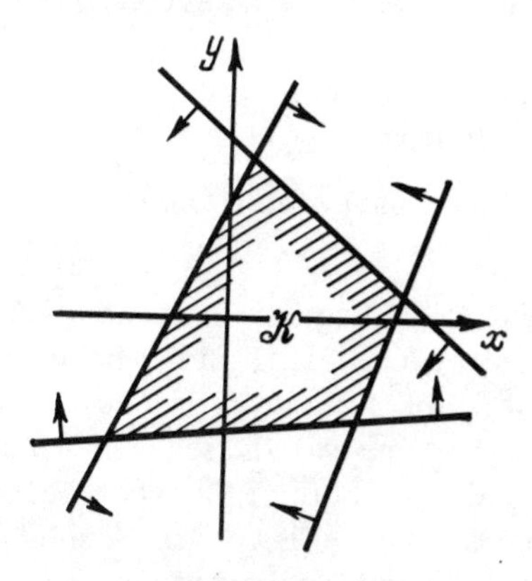

Fig. 2.1

"empty"; in this case we say that the system (2.1) is *inconsistent*. Such a situation is shown in figure 2.3.

A solution domain $\mathscr{K}$ is always *convex*. We recall that by the usual definition, a point set (in the plane or in three-space) is said to be convex if, for any two points A and B in $\mathscr{K}$, the entire line segment AB is contained in $\mathscr{K}$. Figure 2.4 illustrates the difference between convex

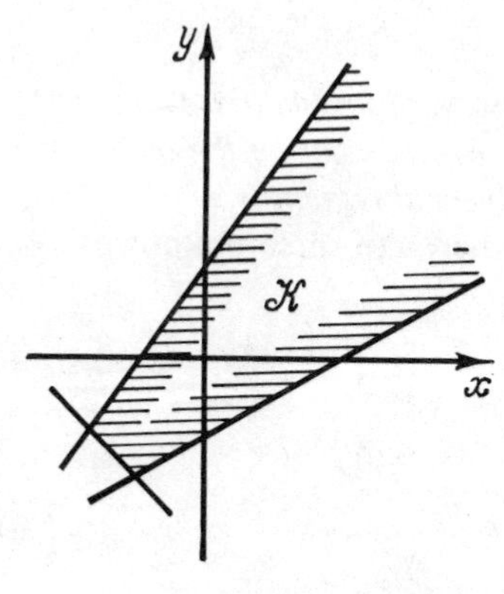

Fig. 2.2

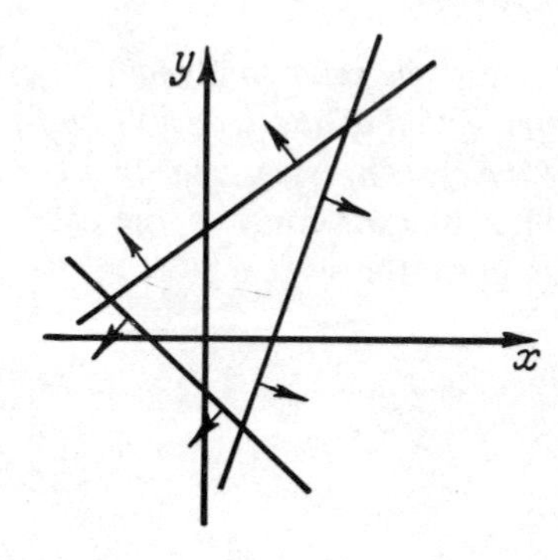

Fig. 2.3

and nonconvex sets. The convexity of a solution domain $\mathscr{K}$ follows from the very means by which the domain is formed; it is obtained as the intersection of several half-planes, and each half-plane is a convex set.

To make clear the fact that solution domains $\mathscr{K}$ are convex, we shall prove the following lemma.

LEMMA 2.1. *The intersection of any finite number of convex sets is a convex set.*

Proof. Let $\mathscr{K}_1$ and $\mathscr{K}_2$ be two convex sets and let $\mathscr{K}$ be their intersection. Consider any two points A and B belonging to $\mathscr{K}$ (fig. 2.5). Since $A \in \mathscr{K}_1$, $B \in \mathscr{K}_1$, and the set $\mathscr{K}_1$ is convex, the segment AB is contained in $\mathscr{K}_1$. In the same way the segment AB is also contained in $\mathscr{K}_2$. Thus, AB is contained in both of the sets $\mathscr{K}_1$ and $\mathscr{K}_2$, and therefore in their intersection $\mathscr{K}$. Thus $\mathscr{K}$ is convex. A similar argument (or an induction) shows that the intersection of any finite number of convex sets is itself convex.

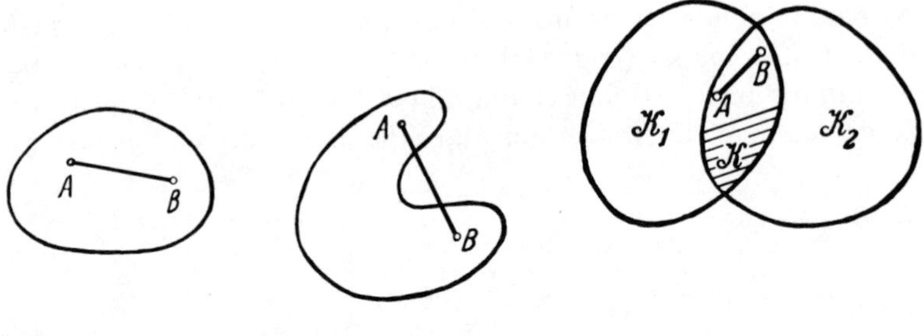

Fig. 2.4

Fig. 2.5

Thus, *the solution domain $\mathscr{K}$ of the system of inequalities* (2.1), *as the intersection of the* (*convex*) *half-planes corresponding to the inequalities of the system, is necessarily a convex polygonal region.*

Let us now consider the case where there are three unknowns. We are given the system

$$
\begin{aligned}
a_1x + b_1y + c_1z + d_1 &\geq 0 \\
a_2x + b_2y + c_2z + d_2 &\geq 0 \\
\cdots\cdots\cdots\cdots&\cdots\cdots \\
a_mx + b_my + c_mz + d_m &\geq 0 .
\end{aligned}
\tag{2.2}
$$

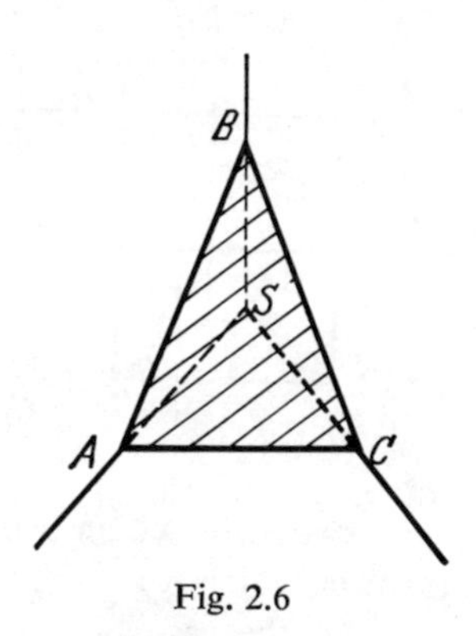

Fig. 2.6

As we know from chapter 1, each of these inequalities defines some half-space. Therefore the region determined by the given system can be represented as the intersection of m half-spaces. However, the intersection of a finite number of half-spaces is some convex polyhedral region $\mathscr{K}$. An example of such a region for the case $m = 4$ is presented in figure 2.6. In this example the region $\mathscr{K}$ is an ordinary tetrahedron. In general, it is not difficult to see that every convex polyhedron is the intersection of a finite number of half-spaces. Of course, it may happen that the region $\mathscr{K}$ is not bounded; an example of such a region is shown in figure 2.7. Finally, it may happen that there are no points at all which satisfy all the inequalities (that is, the system (2.2) is inconsistent); in that case the region $\mathscr{K}$ is empty. Such a situation is represented in figure 2.8.

We must give special consideration to the case where among the inequalities (2.2) there are two of the form

$$ax + by + cz + d \geq 0\,,$$

$$-ax - by - cz - d \geq 0\,,$$

which can be replaced by the single equation

$$ax + by + cz + d = 0\,.$$

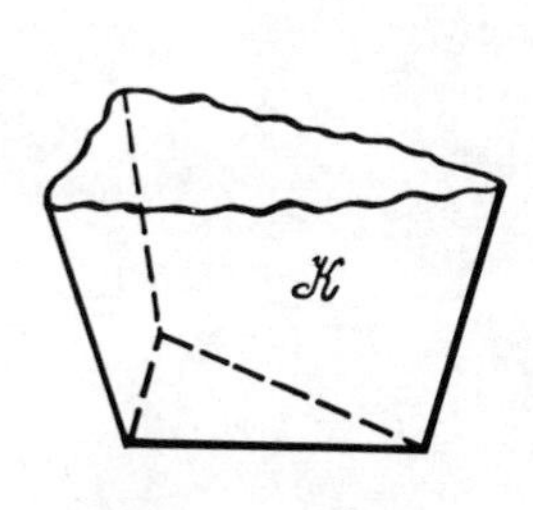

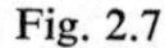

Fig. 2.7

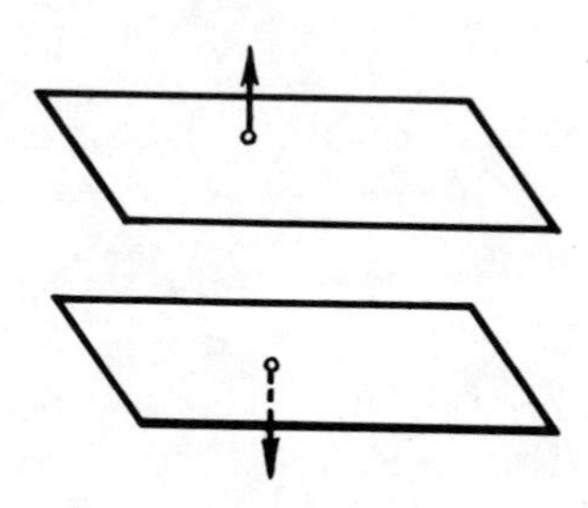

Fig. 2.8

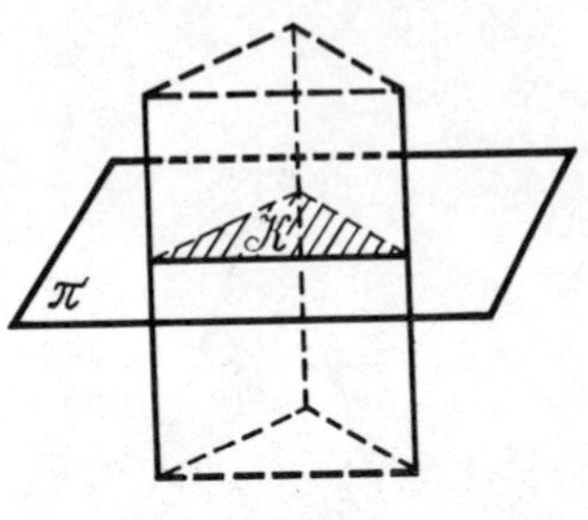

Fig. 2.9

This equation defines a plane π in three-space while the remaining inequalities in (2.2) isolate some convex polygonal region in the plane π as the solution domain of the system (2.2). We see then that *a convex polygonal region in the plane is a special case of a convex polyhedral region in three-space.* In figure 2.9 the region $\mathcal{K}$ is a triangle formed by the intersection of five half-spaces, two of which are bounded by the "horizontal" plane π, and three of which form the "vertical" triangular prism.

As in the case of two unknowns, we shall call the region $\mathcal{K}$ the *solution domain* of the system (2.2). Once again we stress the fact that $\mathcal{K}$, as the intersection of a finite number of half-spaces, is necessarily convex.

Therefore, *the system* (2.2) *determines a convex polyhedral region* $\mathcal{K}$ *in three-space which is the intersection of the half-spaces corresponding to the inequalities of the system.*

If the region $\mathcal{K}$ is bounded, it is called the *solution polyhedron* of the system (2.2).

3 The Convex Hull of a Point Set

Imagine that in the plane, considered as an infinite sheet of paper, nails are inserted at the points $A_1, A_2, \ldots, A_p$. Next, imagine that a rubber band is stretched out into a loop large enough to surround the entire collection of nails (as shown by the dashed line in fig. 3.1), and then allowed to contract as far as the nails will allow. The set of points enclosed by the contracted rubber band is shaded in figure 3.1. This point set obviously forms a convex polygon, called the *convex hull* of the set of points $A_1, A_2, \ldots, A_p$.

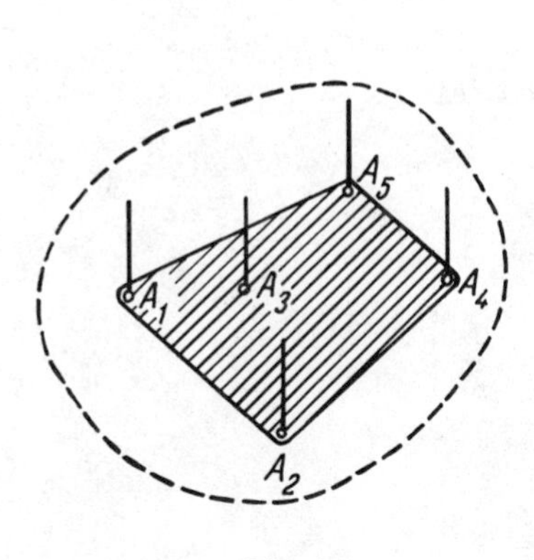

Fig. 3.1

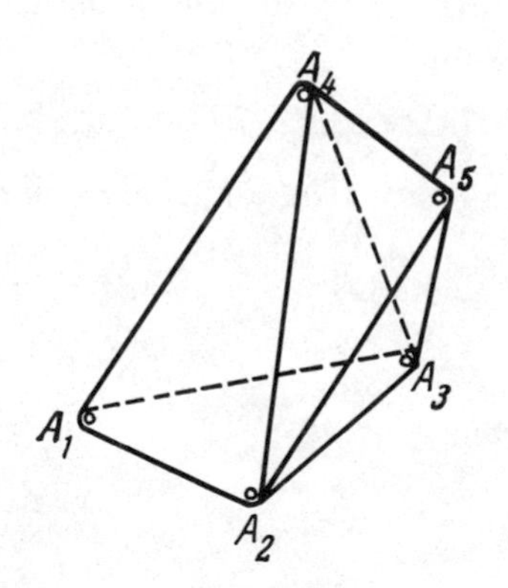

Fig. 3.2

If the points $A_1, A_2, \ldots, A_p$ were distributed over three-space, the analogous experiment would be quite difficult to carry out. But let us give free rein to our imaginations and assume that we have succeeded in surrounding $A_1, A_2, \ldots, A_p$ by a sack made of a rubber sheet. Of its own accord the sack will contract freely until it runs into some of the points. When the contraction has proceeded as far as possible (fig. 3.2), it is clear that the sack will have the form of a convex polyhedron with vertices at several of the points $A_1, A_2, \ldots, A_p$. The region of space enclosed by this polyhedron is again called the *convex hull* of the point set $A_1, A_2, \ldots, A_p$.

Although the above definition of the convex hull is quite clear intuitively, it is not completely flawless from the viewpoint of mathematical rigor, and we shall now define this concept more carefully.

Let $A_1, A_2, \ldots, A_p$ be an arbitrary collection of points (in the plane or in three-space), and consider all points of the form

$$s_1A_1 + s_2A_2 + \cdots + s_pA_p \,, \tag{3.1}$$

where $s_1, s_2, \ldots, s_p$ is any set of nonnegative numbers whose sum is 1:

$$s_1, s_2, \ldots, s_p \geq 0 \quad \text{and} \quad s_1 + s_2 + \cdots + s_p = 1 \,. \tag{3.2}$$

DEFINITION 3.1. *The set of points of the form* (3.1) *for which* (3.2) *is satisfied is called the convex hull of the set of points* $A_1, A_2, \ldots, A_p$, *and is denoted by*

$$\langle A_1, A_2, \ldots, A_p \rangle \,.$$

To justify the use of the term *convex* hull, we prove the following lemma.

LEMMA 3.1. *The set* $\langle A_1, \ldots, A_p \rangle$ *is convex.*

Proof. If M and N are points of $\langle A_1, \ldots, A_p \rangle$, we may write

$$M = s_1A_1 + \cdots + s_pA_p \,,$$
$$N = t_1A_1 + \cdots + t_pA_p \,,$$

where

$$s_1, \ldots, s_p, t_1, \ldots, t_p \geq 0 \,,$$
$$s_1 + \cdots + s_p = t_1 + \cdots + t_p = 1 \,.$$

If u and v are nonnegative numbers, such that $u + v = 1$, then

$$\begin{aligned} uM + vN &= u(s_1A_1 + \cdots + s_pA_p) + v(t_1A_1 + \cdots + t_pA_p) \\ &= (us_1 + vt_1)A_1 + \cdots + (us_p + vt_p)A_p \,. \end{aligned}$$

Since the $us_i + vt_i$ are clearly nonnegative and

$$(us_1 + vt_1) + \cdots + (us_p + vt_p) = u(s_1 + \cdots + s_p) + v(t_1 + \cdots + t_p)$$
$$= u \cdot 1 + v \cdot 1 = u + v = 1 ,$$

the point $uM + vN \in \langle A_1, \ldots, A_p \rangle$ and thus the entire line segment MN is contained in $\langle A_1, \ldots, A_p \rangle$, proving that $\langle A_1, \ldots, A_p \rangle$ is convex.

To convince ourselves that definition 3.1 coincides with the intuitive definition given before, we look first at the cases $p = 2$ and $p = 3$. If $p = 2$, we are given two points A_1 and A_2. As is shown by assertion 1 of chapter 1, the set $\langle A_1, A_2 \rangle$ is the line segment A_1A_2.

If $p = 3$, we have three points A_1, A_2 and A_3. We shall show that the set $\langle A_1, A_2, A_3 \rangle$ consists of all the points lying inside and on the edges of the triangle $A_1A_2A_3$.

For the general case we shall prove the following lemma:

LEMMA 3.2. *The set $\langle A_1, \ldots, A_{p-1}, A_p \rangle$ is the union of all possible line segments connecting the points A_p to points of the set $\langle A_1, \ldots, A_{p-1} \rangle$.*

Proof. For simplicity of notation we shall denote the set $\langle A_1, \ldots, A_{p-1} \rangle$ by $\mathscr{M}_{p-1}$, and the set $\langle A_1, \ldots, A_{p-1}, A_p \rangle$ by $\mathscr{M}_p$.

Let us consider any point $A \in \mathscr{M}_p$. A necessarily has the form

$$A = s_1A_1 + \cdots + s_{p-1}A_{p-1} + s_pA_p ,$$

where

$$s_1, \ldots, s_p \geq 0, \qquad s_1 + \cdots + s_p = 1 .$$

If $s_p = 0$, then $A \in \mathscr{M}_{p-1}$; thus the set $\mathscr{M}_{p-1}$ is contained in $\mathscr{M}_p$. If $s_p = 1$, then $A = A_p$; hence the point A_p belongs to $\mathscr{M}_p$. By lemma 3.1, $\mathscr{M}_p$ is convex; thus all possible line segments $A'A_p$ connecting points $A' \in \mathscr{M}_{p-1}$ to A_p are entirely contained in $\mathscr{M}_p$.

Now we need only verify that the set $\mathscr{M}_p$ contains no more than the union of these segments; that is, that each $A \in \mathscr{M}_p$ lies on at least one of these segments.

Let $A \in \mathscr{M}_p$; then A is of the form (3.1), and (3.2) is satisfied. We may assume that $s_p \neq 1$, for otherwise we would have $A = A_p$ and there would be nothing to prove. However, if $s_p \neq 1$, then $s_p < 1$, and $s_1 + \cdots + s_{p-1} = 1 - s_p > 0$, and we may write

$$A = (s_1 + \cdots + s_{p-1}) \left[\frac{s_1}{s_1 + \cdots + s_{p-1}} A_1 + \cdots \right.$$
$$\left. + \frac{s_{p-1}}{s_1 + \cdots + s_{p-1}} A_{p-1} \right] + s_pA_p .$$

The expression in square brackets defines a point $A' \in \mathcal{M}_{p-1}$, since the coefficients of $A_1, \ldots, A_{p-1}$ in this expression are nonnegative and add up to 1. Thus, our equation becomes

$$A = (s_1 + \cdots + s_{p-1})A' + s_p A_p .$$

Since the coefficients of A' and A_p are nonnegative and add up to 1, the point A lies on the line segment $A'A_p$. Thus the proof of the lemma is complete.

It is now a simple matter to see that the intuitive definition of a convex hull given at the beginning of this section is equivalent to the rigorous definition given above. Whichever of the two definitions of the convex hull we take, we move from the convex hull of the system $A_1, \ldots, A_{p-1}$ to that of the system $A_1, \ldots, A_{p-1}, A_p$ according to the same rule: we construct segments connecting the point A_p to all points of the convex hull of the points $A_1, \ldots, A_{p-1}$. (For the intuitive definition, this rule is obvious; for the rigorous definition, it is formulated in lemma 3.2.) If we now note that for $p = 2$ we obtain the same set according to both definitions—the line segment A_1A_2—then the equivalence of the two definitions is clear (by induction on p).

In addition, it is easy to see that $\langle A_1, A_2, \ldots, A_p \rangle$ is not only a convex set, but the "smallest" convex set containing the points $A_1, A_2, \ldots, A_p$, in the sense that it is contained in every such set. This statement follows immediately from lemma 3.2 and from the definition of a convex set.

The fact that $\langle A_1, A_2, \ldots, A_p \rangle$ is the smallest convex set containing the points $A_1, A_2, \ldots, A_p$ reveals the motivation for the intuitive rubber band (or rubber sheet) definition given at the beginning of this section: the set enclosed by the rubber band (or sheet), after it has contracted as closely as possible about the points $A_1, A_2, \ldots, A_p$, is certainly the smallest convex set containing these points.

4 Convex Polyhedral Cones

We begin with a definition.

DEFINITION 4.1. *A convex polyhedral cone is the intersection of a finite number of half-spaces whose bounding planes all pass through a common point, called the vertex of the cone.*

We shall first discuss the relationship between this concept and systems of linear inequalities. Let us consider the special case in which the vertex of the cone lies at the origin—that is, in which all the boundary planes pass through the origin. The equations of such planes are of the form

$$ax + by + cz = 0$$

(the constant term of the equation must be zero, otherwise (0, 0, 0) would not be a solution). Therefore *a convex polyhedral cone with vertex at the origin is the solution domain of some system of homogeneous linear inequalities* (that is, linear inequalities with constant term 0);

$$\begin{aligned} a_1x + b_1y + c_1z &\geq 0 \\ a_2x + b_2y + c_2z &\geq 0 \\ \cdots\cdots\cdots&\cdots\cdots \\ a_mx + b_my + c_mz &\geq 0 . \end{aligned}$$

The converse, of course, is also true: the solution domain of a system of homogeneous linear inequalities is always a convex polyhedral cone with vertex at the origin.

As an example of a convex polyhedral cone, we consider the convex region in space whose boundary is a polyhedral angle with vertex S,

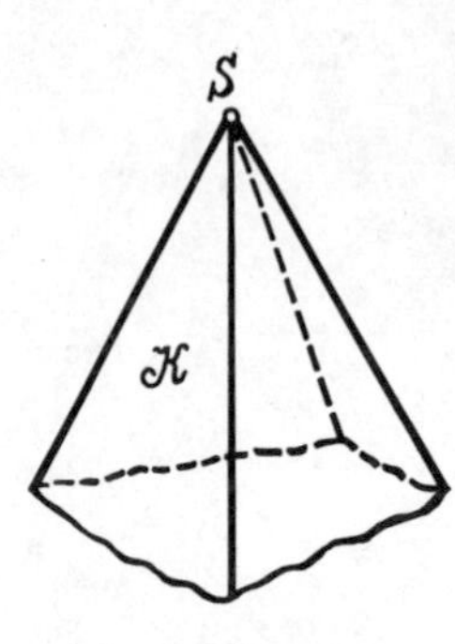

Fig. 4.1

somewhat like an infinite convex pyramid which has no base and extends infinitely far away from the vertex (a four-sided pyramid of this kind is sketched in fig. 4.1). Less interesting examples can also be cited:

1. A half-space (fig. 4.2*a*). For such a cone the vertex could be any point $S \in \pi$, where π is the bounding plane.

2. The intersection of two half-spaces whose bounding planes intersect in some line l (fig. 4.2*b*). Any point $S \in l$ can be the vertex.

3. A plane. It is clear that any plane π in three-space can be considered as the intersection of the two half-spaces bounded by π (fig. 4.2*c*). In this case, the vertex could be any point $S \in \pi$.

4. A half-plane (fig. 4.2*d*). The vertex can be any point on the boundary.

5. A straight line. Every straight line l in three-space can be considered as the intersection of three half-spaces whose bounding planes intersect in the line l (fig. 4.2*e*). The vertex can be any point on the line l.

6. An angular sector (smaller than 180°) in any plane π (fig. 4.2*f*).

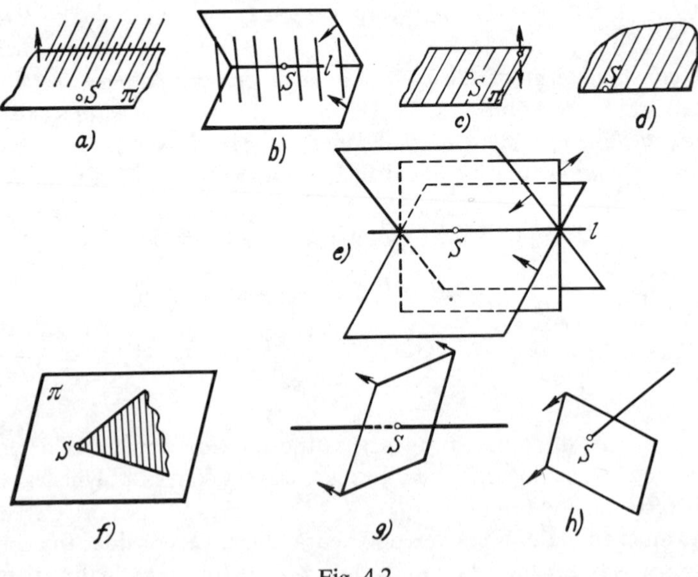

Fig. 4.2

Such a sector can be obtained as the intersection of the plane π with two half-spaces. (How?)

7. A ray (fig. 4.2*g*). A ray can be regarded as the intersection of a line and a half-space. The vertex S is the endpoint of the ray.

8. A point. Such a "cone" can be obtained as the intersection of a ray with a half-space (fig. 4.2*h*).

Of course, examples 1–8 do not conform to our usual notion of a "cone," but they are the degenerate cases that must be accepted if we are to have a truly general and mathematically rigorous definition of a convex polyhedral cone.

We shall now prove that the examples listed above (1–8 and the infinite convex pyramid) are the *only* possibilities for convex polyhedral cones in three-space.

Let $\mathscr{K}$ be a given convex polyhedral cone which is the intersection of p half-spaces. If $p = 1$, our assertion is correct, for then $\mathscr{K}$ is a half-space. A simple argument, the details of which we leave to the reader, shows that if our assertion is true for cones formed as intersections of p half-spaces, then it is also valid for cones formed as intersections of $p + 1$ half-spaces. Hence, by the principle of mathematical induction, it follows that the assertion is true for any p.

Convex polyhedral cones possess a number of interesting properties which, because of the size of this booklet, cannot be examined here in great detail. Nevertheless, we shall attempt to present a number of these features—some in this chapter, and some in chapter 9. But first, we must introduce some notation.

Let $B_1, B_2, \ldots, B_q$ be points in three-space. *We define the symbol* $(B_1, B_2, \ldots, B_q)$ *as the set of points of the form*

$$t_1B_1 + t_2B_2 + \cdots + t_qB_q\,,$$

where $t_1, t_2, \ldots, t_q$ *are arbitrary nonnegative real numbers.*

What geometric object is represented by the set $(B_1, B_2, \ldots, B_q)$? It is clear from the definition that it is the sum of the set $(B_1), (B_2), \ldots, (B_q)$; therefore we should first determine what the set (B) represents. Since (B) is the set of points of the form tB, where t is any nonnegative real number and B is a fixed point, it is easy to see that if B is the origin, then the set (B) consists of the origin; otherwise (B) is the ray with endpoint at the origin which passes through the point B. We now note that the sum of any point set and the set consisting of the origin is precisely the same point set, so it is clear that in speaking of the set $(B_1, B_2, \ldots, B_q)$, we lose no generality by assuming that *none of the points* $B_1, B_2, \ldots, B_q$ *is the origin. Under this assumption, then, the set* $(B_1, B_2, \ldots, B_q)$ *is precisely the sum of the rays* $(B_1), (B_2), \ldots, (B_q)$.

The last remark makes the following lemma almost obvious.

LEMMA 4.1. *The set* $(B_1, \ldots, B_{q-1}, B_q)$ *is the union of all possible line segments joining points of the set* $(B_1, \ldots, B_{q-1})$ *with points of the ray* (B_q).

A rigorous proof of this lemma can be constructed along the lines of the proof of lemma 3.2 after the analogue of lemma 3.1 is proved; we recommend that the reader write out such an argument.

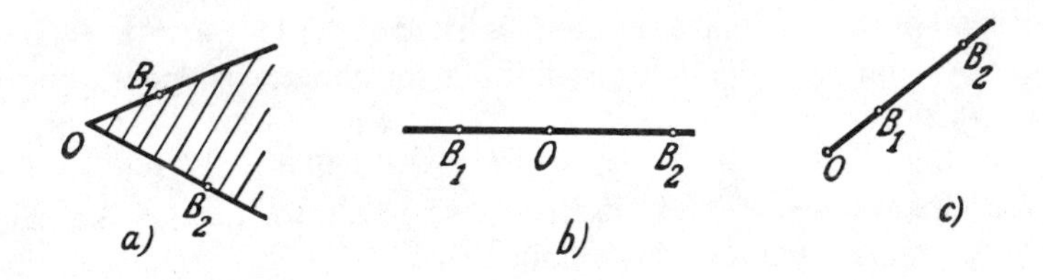

Fig. 4.3

Using lemma 4.1 it is easy to show that (B_1, B_2) (under the assumption that neither B_1 nor B_2 is the origin) is an angular sector, a straight line, or a ray (figs. 4.3*a*, *b*, *c*). Knowing this, it is easily demonstrated that (B_1, B_2, B_3) is an infinite triangular pyramid, a plane, a half-plane, an angular sector, a straight line, or a ray. It is now clear that there is a close connection between sets of the form $(B_1, B_2, \ldots, B_q)$ and convex polyhedral cones. This relationship is formulated in the following two theorems.

THEOREM 4.1. *The set* $(B_1, B_2, \ldots, B_q)$ *either comprises all space or is a convex polyhedral cone with vertex at the origin.*

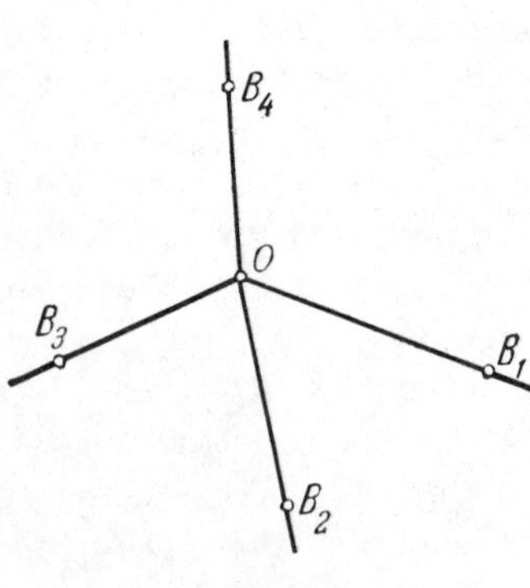

Fig. 4.4

The fact that the set $(B_1, B_2, \ldots, B_q)$ may indeed be all space can be shown by an example. Consider four points B_1, B_2, B_3, B_4 arranged so that the rays (B_1), (B_2), (B_3), (B_4) meet pairwise at obtuse angles (fig. 4.4). Each of the sets (B_1, B_2, B_3), (B_1, B_2, B_4), (B_1, B_3, B_4), and (B_2, B_3, B_4) is an infinite triangular pyramid with vertex at the origin, and it is evident that the set (B_1, B_2, B_3, B_4) contains each of these pyramids. However, the union of these pyramids is all of space.

THEOREM 4.2. *Every convex polyhedral cone with vertex at the origin is a set of the form* $(B_1, B_2, \ldots, B_q)$.

Proof of Theorem 4.1. We shall outline a proof by induction on q. The assertion of the theorem for the case $q = 1$ is obvious. Let us assume that the theorem is true for sets of the form $(B_1, \ldots, B_q)$, and consider sets of the form $(B_1, \ldots, B_q, B_{q+1})$.

According to the inductive hypothesis, $(B_1, \ldots, B_q)$ is either all of space or some convex polyhedral cone. In the first case there is nothing to prove, for then $(B_1, \ldots, B_q, B_{q+1})$ is also all of space. Consider the second case, where $(B_1, \ldots, B_q)$ is a convex polyhedral cone $\mathscr{K}$. According to lemma 4.1, the set $(B_1, \ldots, B_q, B_{q+1})$ is the union of all segments connecting points of $\mathscr{K}$ with points of the ray (B_{q+1}). However, as shown earlier, every convex polyhedral cone $\mathscr{K}$ is either an infinite convex pyramid or a set of one of the forms 1–8. By forming the appropriate union of segments for each of these cases (the reader should do this as an exercise), we find that we obtain either all space or a convex polyhedral cone. Hence the theorem is valid for sets of the form (B_1), and it is also valid for $(B_1, \ldots, B_q, B_{q+1})$ if it holds for $(B_1, \ldots, B_q)$. Therefore it follows that the theorem is true for all q.

Proof of theorem 4.2. Let $\mathscr{K}$ be a convex polyhedral cone with vertex at the origin O. Then $\mathscr{K}$ is either an infinite convex pyramid or one of the sets 1–8.

Suppose $\mathscr{K}$ is an infinite convex pyramid. If we select one point from each of its edges, we have a set of points $B_1, B_2, \ldots, B_q$. We claim that *the set* $(B_1, B_2, \ldots, B_q)$ *is precisely* $\mathscr{K}$.

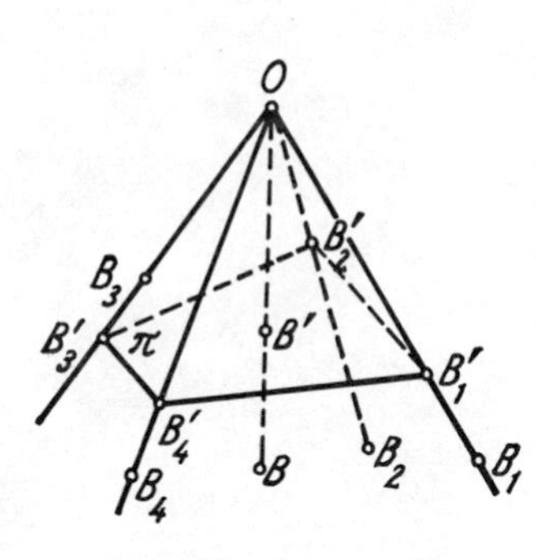

Fig. 4.5

To prove this, consider any plane π intersecting all the edges of the pyramid $\mathscr{K}$. This plane determines points of intersection $B_1', B_2', \ldots, B_q'$ with the edges (fig. 4.5). It is clear that

$$B_1' = k_1 B_1, \quad B_2' = k_2 B_2, \ldots, B_q' = k_q B_q, \tag{4.1}$$

for some nonnegative numbers $k_1, k_2, \ldots, k_q$.

Now let B be any point of the pyramid distinct from the vertex O, and let B' be the point of intersection of the plane π with the ray OB. It is clear that B' belongs to the convex hull of the point set $B_1', B_2', \ldots, B_q'$, and that therefore

$$B' = s_1B'_1 + s_2B'_2 + \cdots + s_qB'_q\,,$$

where $s_1, s_2, \ldots, s_q$ are nonnegative numbers adding up to 1. From equations (4.1) we obtain

$$B' = s_1k_1B_1 + s_2k_2B_2 + \cdots + s_qk_qB_q\,,$$

and if we note that $B' = kB$ $(k > 0)$, we find that

$$B = t_1B_1 + t_2B_2 + \cdots + t_qB_q\,,$$

where $t_i = s_ik_i/k$ $(i = 1, 2, \ldots, q)$. We have therefore shown that every point B of the pyramid belongs to the set $(B_1, B_2, \ldots, B_q)$. The converse (that is, the fact that the set $(B_1, B_2, \ldots, B_q)$ is contained in $\mathscr{K}$) is obvious. Thus, $\mathscr{K}$ is identical to $(B_1, B_2, \ldots, B_q)$.

The cases in which $\mathscr{K}$ is one of the exceptional sets 1–8 can be treated without any difficulty; they are left to the reader.

5 The Solution Domain of a System of Inequalities in Two Unknowns

Our task now is to provide a description of all the solutions of a given system of linear inequalities. In this chapter, the problem is solved for the case of systems in two unknowns x and y. However, we shall attempt to carry out our analysis with as much generality as possible so that it can be easily extended to a larger number of unknowns.

The solution of any system of linear inequalities ultimately reduces to the solution of a system of linear equations. We shall consider the solution of systems of linear equations as a simple procedure, and we shall not be disturbed if we are required to do this a large number of times.

5.1. Necessary Lemmas

Consider the system of inequalities

$$\begin{gathered} a_1x + b_1y + c_1 \geq 0 \\ a_2x + b_2y + c_2 \geq 0 \\ \cdots\cdots\cdots\cdots\cdots\cdots \\ a_mx + b_my + c_m \geq 0\,. \end{gathered} \tag{5.1}$$

It will expedite matters to consider at the same time the corresponding system of homogeneous inequalities

$$\begin{gathered} a_1x + b_1y \geq 0 \\ a_2x + b_2y \geq 0 \\ \cdots\cdots\cdots\cdots \\ a_mx + b_my \geq 0 \end{gathered} \tag{5.2}$$

and the corresponding system of homogeneous equations

$$\begin{aligned} a_1x + b_1y &= 0 \\ a_2x + b_2y &= 0 \\ &\cdots\cdots \\ a_mx + b_my &= 0\,. \end{aligned} \tag{5.3}$$

We shall denote the solution domain of (5.1) in the xy coordinate plane by $\mathscr{K}$, that of the system (5.2) by $\mathscr{K}_0$, and that of the system (5.3) by $\mathscr{L}$. It is clear that $\mathscr{L} \subset \mathscr{K}_0$, where the symbol $\subset$ is an abbreviation for the words "is contained in."

LEMMA 5.1.

$$\mathscr{K} + \mathscr{K}_0 \subset \mathscr{K}\,;$$

that is, the sum of any solution of a given system of inequalities with any solution of the corresponding system of homogeneous inequalities is again a solution of the original system.

Proof. Let $A = (x_A, y_A)$ be any point of $\mathscr{K}$, and let $B = (x_B, y_B)$ be any point of $\mathscr{K}_0$. Then we have the inequalities

$$\begin{aligned} a_1x_A + b_1y_A + c_1 &\geq 0 \\ a_2x_A + b_2y_A + c_2 &\geq 0 \\ &\cdots\cdots \\ a_mx_A + b_my_A + c_m &\geq 0 \end{aligned} \qquad \text{and} \qquad \begin{aligned} a_1x_B + b_1y_B &\geq 0 \\ a_2x_B + b_2y_B &\geq 0 \\ &\cdots\cdots \\ a_mx_B + b_my_B &\geq 0\,. \end{aligned}$$

By adding each inequality on the right side to the corresponding inequality on the left, we obtain

$$\begin{aligned} a_1(x_A + x_B) + b_1(y_A + y_B) + c_1 &\geq 0 \\ a_2(x_A + x_B) + b_2(y_A + y_B) + c_2 &\geq 0 \\ &\cdots\cdots \\ a_m(x_A + x_B) + b_m(y_A + y_B) + c_m &\geq 0\,, \end{aligned}$$

implying that the point $A + B = (x_A + x_B, y_A + y_B)$ is a solution to the original system (5.1) of inequalities; that is, that $A + B \in \mathscr{K}$.

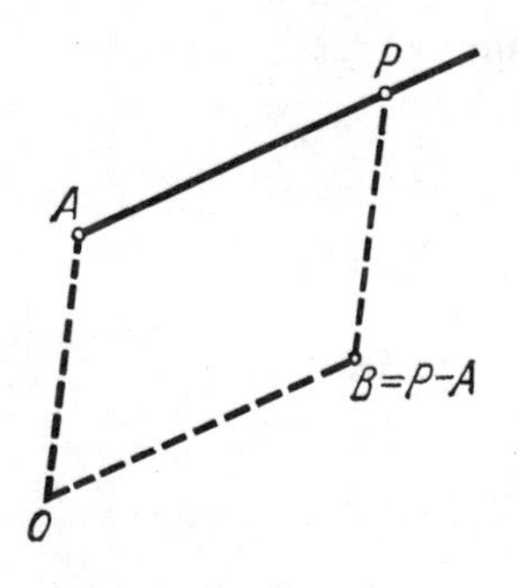

Fig. 5.1

LEMMA 5.2. (*a*) *If a ray with initial point A is entirely contained in the set* $\mathscr{K}$ *and P is any point on this ray, then* $P - A \in \mathscr{K}_0$. (*b*) *If a straight line belongs entirely to the set* $\mathscr{K}$ *and A and P are any two points of this line, then* $P - A \in \mathscr{L}$.

1. *Proof of lemma* 5.2(*a*). Denote the point $P - A$ by B. The ray under consideration then consists of points of the form

$$A + sB = (x_A + sx_B, y_A + sy_B)\,, \quad (5.4)$$

where s is any nonnegative number (fig. 5.1). As a solution of the system (5.1), each such point satisfies the inequalities

$$\begin{aligned} a_1(x_A + sx_B) + b_1(y_A + sy_B) + c_1 &\geq 0 \\ a_2(x_A + sx_B) + b_2(y_A + sy_B) + c_2 &\geq 0 \\ \cdots\cdots\cdots\cdots\cdots\cdots\cdots\cdots\cdots& \\ a_m(x_A + sx_B) + b_m(y_A + sy_B) + c_m &\geq 0\,. \end{aligned} \quad (5.5)$$

Consider, for example, the first of these inequalities, which we may write in the form

$$(a_1x_A + b_1y_A + c_1) + s(a_1x_B + b_1y_B) \geq 0\,.$$

Since this inequality is valid for *every* $s \geq 0$, it is not difficult to see that the coefficient of s must be nonnegative:

$$a_1x_B + b_1y_B \geq 0\,.$$

Similarly, by considering the remaining inequalities in (5.5), we find that

$$\begin{aligned} a_2x_B + b_2y_B &\geq 0 \\ \cdots\cdots\cdots\cdots& \\ a_mx_B + b_my_B &\geq 0\,, \end{aligned}$$

from which it is clear that the point B belongs to the set $\mathscr{K}_0$.

2. The proof of lemma 5.2(*b*) is similar. The straight line is composed of points of the form (5.4), where s runs over all real numbers. Therefore the inequalities (5.5) are satisfied for any s. It follows, therefore, that

in each of these inequalities the coefficient of s must be zero; that is, that

$$a_1x_B + b_1y_B = 0$$

$$a_2x_B + b_2y_B = 0$$

$$\dots\dots\dots\dots\dots$$

$$a_mx_B + b_my_B = 0\,.$$

Consequently, $B \in \mathscr{L}$, and the lemma is proved.

It is easy to see that lemmas 5.1 and 5.2 remain valid for systems in any numbers of unknowns.

5.2. The Case of a Normal System of Inequalities

Once again consider the system of inequalities (5.1) and the corresponding system of homogeneous equations (5.3). The latter system has the trivial solution $x = 0$, $y = 0$, which is called the *zero* solution. To investigate the solution domain of (5.1), it is important to determine whether (5.3) has any nonzero solutions. In this connection we make the following definition.

DEFINITION 5.1. *A system of linear inequalities is called normal if the corresponding system of homogeneous equations has only the zero solution.*

In other words, a system of inequalities is normal if the solution domain of the corresponding system of homogeneous equations contains only one point (the origin). Since each homogeneous equation determines a straight line passing through the origin, this means simply that not all of these lines are the same; that is, that the system of homogeneous equations cannot be replaced by a single homogeneous equation.

The concept of a normal system can be defined in the same way for any number of unknowns. The geometric impact of the concept of normality is embodied in the following theorem.

THEOREM 5.1. *A consistent system of linear inequalities is normal if and only if its solution domain $\mathscr{K}$ contains no straight line.*

Proof. If the system is normal, that is, if the set $\mathscr{L}$ contains only the origin, then the region $\mathscr{K}$ contains no lines; this follows immediately from the second assertion of lemma 5.2. If the system is not normal, then the set $\mathscr{L}$ contains at least one point B different from the origin. Then, of course, all points of the form kB, where k is any real number, also belong to $\mathscr{L}$.[1] In this case, however, for any point $P \in \mathscr{K}$ (such a point must exist since the system is consistent), the set of all points of the form $P + kB$ (where k is any real number) is contained in $\mathscr{K}$ by lemma 5.1. As we know, such a set is a straight line. Thus, when the system is not normal, the region $\mathscr{K}$ contains a line. This completes the proof of the theorem.

In this section we shall examine the solution domain for the system (5.1) under the assumption that it is consistent (that is, that the region $\mathscr{K}$ is not empty) and normal.

First of all, from the condition that the region $\mathscr{K}$ is normal (and thus contains no lines), it follows that $\mathscr{K}$ must have vertices. We give the word "vertex" the formal interpretation (which is very close to the intuitive one) as shown in figure 5.2.

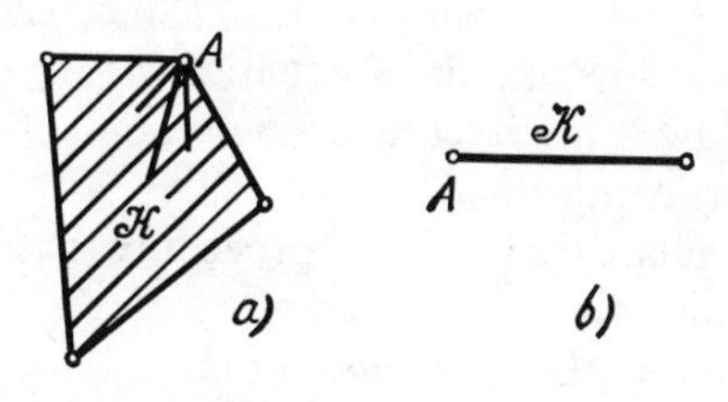

Fig. 5.2

DEFINITION 5.2. *A vertex of a region $\mathscr{K}$ is a point of $\mathscr{K}$ which is not an interior point of any line segment lying entirely within $\mathscr{K}$.*

In other words, a vertex is a point $A \in \mathscr{K}$ which has the property that any line segment passing through A which is contained in $\mathscr{K}$ must have A as an endpoint (fig. 5.2*a*, *b*).

Let us explain in more detail why the convex region $\mathscr{K}$ under consideration necessarily has vertices. If $\mathscr{K}$ lies on a straight line, it is either a single point, a segment, or a ray. If, however, $\mathscr{K}$ does not lie on a line, then its boundary must consist of segments and/or rays (since $\mathscr{K}$ does not contain lines). It is clear that the endpoint of any such segment or ray must be a vertex of $\mathscr{K}$.

Finding the vertices of a region $\mathscr{K}$ presents no difficulty. First we note

1. If the coordinates of the point $B = (x, y)$ satisfy a system of homogeneous equations, then the coordinates of the point $kB = (kx, ky)$ also satisfy that system.

that the ith inequality of (5.1) corresponds to a half-plane whose boundary is the straight line l_i defined by the equation

$$a_i x + b_i y + c_i = 0 .$$

It is clear that a point A of $\mathscr{K}$ is a vertex if and only if it belongs to *two* such boundary lines.

We shall say that any pair of equations from the nonhomogeneous system

$$\begin{aligned} a_1 x + b_1 y + c_1 &= 0 , \\ a_2 x + b_2 y + c_2 &= 0 , \\ \dots\dots\dots\dots &\dots , \\ a_m x + b_m y + c_m &= 0 . \end{aligned} \tag{5.6}$$

is *regular* if it has a *unique* solution (x, y). Since each such equation determines a straight line, a pair of equations is regular if and only if the corresponding lines intersect at a single point; that is, if they are neither parallel nor identical.

We can now formulate the following procedure for finding the vertices of a solution domain $\mathscr{K}$.

In order to find all vertices, we solve all the regular two-equation subsystems of (5.6) *and discard those points which do not satisfy the original system* (5.1).

Since the number of such regular subsystems cannot exceed the total number of two-equation subsystems of (5.6) (the binomial coefficient $\binom{m}{2}$, which is the number of ways of selecting two out of m objects), $\mathscr{K}$ has at most $\binom{m}{2}$ vertices. In particular, *the number of vertices of $\mathscr{K}$ is finite.*

REMARK 5.1. It follows that if the solution domain $\mathscr{K}$ of a normal system has no vertices, then the domain is empty and the system has no solutions (is inconsistent).

EXAMPLE 5.1. Find all vertices of the solution domain $\mathscr{K}$ for the system

$$\begin{aligned} x + y + 1 &\geq 0 \\ x - 2y - 2 &\geq 0 \\ 2x - y - 4 &\geq 0 . \end{aligned}$$

By solving the subsystems

$$\text{(i)}\ \begin{aligned} x + y + 1 &= 0 \\ x - 2y - 2 &= 0, \end{aligned} \quad \text{(ii)}\ \begin{aligned} x + y + 1 &= 0 \\ 2x - y - 4 &= 0, \end{aligned} \quad \text{(iii)}\ \begin{aligned} x - 2y - 2 &= 0 \\ 2x - y - 4 &= 0, \end{aligned}$$

all of which are regular, we obtain the three points

$$(0, -1), (1, -2), (2, 0),$$

of which only the second and third satisfy all the inequalities. This means that the vertices of $\mathscr{K}$ are the points

$$A_1 = (1, -2) \quad \text{and} \quad A_2 = (2, 0).$$

Returning to the system (5.1), the following theorem will allow us to use our method of calculating the vertices of the region $\mathscr{K}$ to find a method of calculating $\mathscr{K}$ itself.

THEOREM 5.2. *If the system* (5.1) *is normal and* $A_1, \ldots, A_p$ *are the vertices of its solution domain* $\mathscr{K}$, *then*

$$\mathscr{K} = \langle A_1, A_2, \ldots, A_p \rangle + \mathscr{K}_0. \tag{5.7}$$

Proof. $\mathscr{K}$ is a convex set containing the points $A_1, \ldots, A_p$, and thus $\mathscr{K}$ contains the "smallest" convex set containing these points; that is, $\langle A_1, A_2, \ldots, A_p \rangle \subset \mathscr{K}$. Thus, by lemma 5.1,

$$\langle A_1, A_2, \ldots, A_p \rangle + \mathscr{K}_0 \subset \mathscr{K}.$$

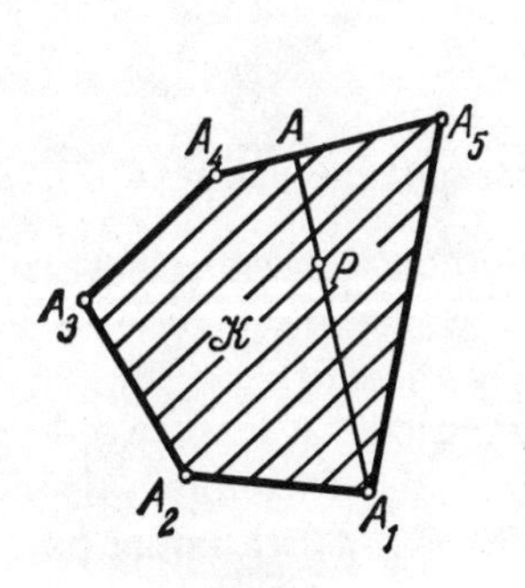

Fig. 5.3

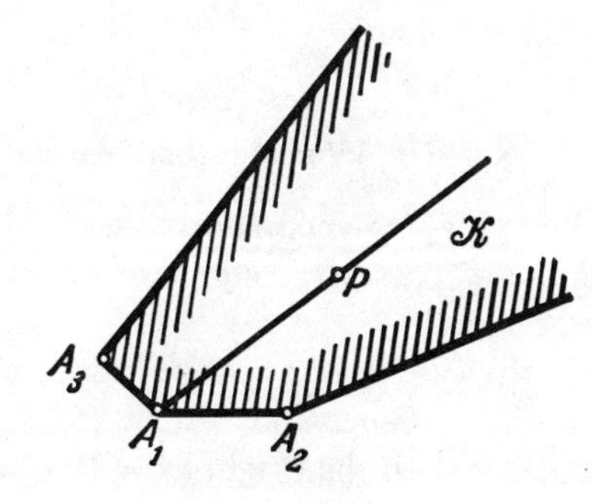

Fig. 5.4

To prove inclusion in the other direction, let P be a point of $\mathscr{K}$ distinct from the vertices. (If such a point does not exist, the inclusion is trivial.) The straight line A_1P intersects the convex region $\mathscr{K}$ either along a line segment A_1A (fig. 5.3), or along a ray with endpoint A_1 (fig. 5.4).

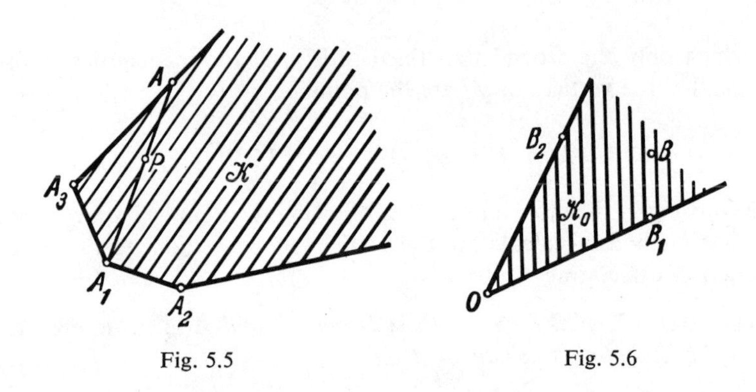

Fig. 5.5 Fig. 5.6

In the second case $P - A_1 \in \mathscr{K}_0$ (by lemma 5.2), so that $P \in A_1 + \mathscr{K}_0$. In the first case we reason as follows: if A lies on an edge A_iA_j of the region $\mathscr{K}$ (as in fig. 5.3), then P belongs to the convex hull of the points A_1, A_i, and A_j; if A lies on an unbounded edge starting from one of the vertices A_i (fig. 5.5), then by lemma 5.2 we have $A \in A_i + \mathscr{K}_0$, implying that $P \in \langle A_1, A_i \rangle + \mathscr{K}_0$. Hence, in each case the point P belongs to the set $\langle A_1, A_2, \ldots, A_p \rangle + \mathscr{K}_0$, and the theorem is proved.

Since we already have a procedure for finding the vertices, we lack only a way of finding $\mathscr{K}_0$ (the solution domain of the homogeneous system (5.2)) to have a complete description of $\mathscr{K}$. We now turn to this problem.

5.3. The Homogeneous Normal System of Inequalities (5.2)

Each of the inequalities in (5.2) determines a half-plane whose boundary passes through the origin; $\mathscr{K}_0$ is the intersection of these half-planes.

In the case under consideration, there must be at least two distinct boundaries (since the system (5.2) is normal). Therefore, $\mathscr{K}_0$ either coincides with the origin ($x = 0, y = 0$), is a ray with initial point at the origin, or is an angular sector, smaller than 180°, with vertex at the origin. If we could find two points B_1 and B_2 lying on different edges of this angle (fig. 5.6), then the entire sector could be described as the set of points of the form

$$B = t_1 B_1 + t_2 B_2 , \tag{5.8}$$

where t_1 and t_2 are arbitrary nonnegative numbers. But finding such points is not particularly difficult once we realize that each of them (*a*) belongs to $\mathscr{K}_0$, that is, satisfies the system (5.2); and (*b*) lies on the boundary of $\mathscr{K}_0$, that is, satisfies one of the equations (5.3). If $\mathscr{K}_0$ is a ray, then instead of equation (5.8) we have

$$B = tB_1 , \tag{5.9}$$

where B_1 is any point on this ray (apart from the origin), and t is any nonnegative number.

EXAMPLE 5.2. Find the solution domain $\mathscr{K}_0$ of the system

$$\begin{aligned} x + y &\geq 0 \\ x - 2y &\geq 0 \\ 2x - y &\geq 0 , \end{aligned} \tag{5.10}$$

and the solution domain $\mathscr{K}$ of the system of example 5.1.

Solution. The system (5.10) is normal, and the unique solution of the corresponding homogeneous equations

$$\begin{aligned} x + y &= 0 \\ x - 2y &= 0 \\ 2x - y &= 0 \end{aligned} \tag{5.11}$$

is (0, 0).

Take any point (except the origin) satisfying the first of the equations (5.11), for example, the point $C = (-1, 1)$. A simple check convinces us that the point C does not satisfy all the inequalities in (5.10), so that neither it nor any point on the ray OC (apart from the origin) belongs to $\mathscr{K}_0$. We find, however, that the point $-C = (1, -1)$ belongs to $\mathscr{K}_0$, so that we may pick $B_1 = -C$. The point (2, 1) satisfies the second equation and is also a solution of the system (5.10), so that we

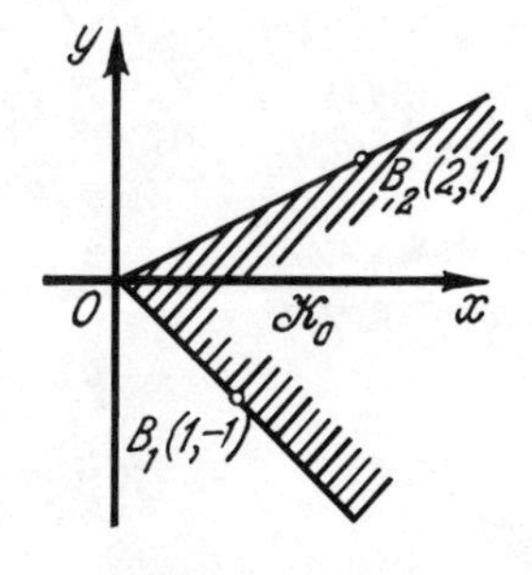

Fig. 5.7

may pick $B_2 = (2, 1)$. The region $\mathscr{K}_0$ thus consists of the points (fig. 5.7)

$$t_1B_1 + t_2B_2 = t_1(1, -1) + t_2(2, 1) = (t_1 + 2t_2, -t_1 + t_2),$$

where t_1 and t_2 are any nonnegative numbers.

Turning to the system of inequalities in example 5.1, we see that the corresponding system of homogeneous inequalities is precisely (5.10). By theorem 5.2, we have

$$\mathscr{K} = \langle A_1, A_2 \rangle + \mathscr{K}_0,$$

where $A_1 = (1, -2)$ and $A_2 = (2, 0)$ are the vertices of the region $\mathscr{K}$. Therefore $\mathscr{K}$ is the set of points of the form (fig. 5.8)

$$\begin{aligned} s(1, -2) + (1 - s)(2, 0) + (t_1 + 2t_2, -t_1 + t_2) \\ = (2 - s + t_1 + 2t_2, -2s - t_1 + t_2), \end{aligned}$$

where s is any number from 0 to 1 inclusive, and t_1 and t_2 are any nonnegative numbers.

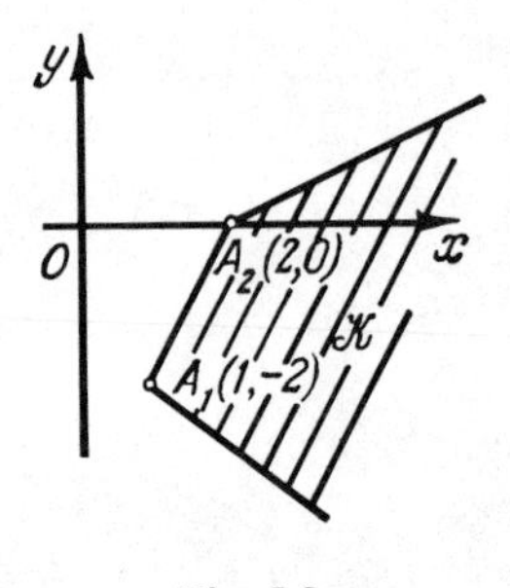

Fig. 5.8

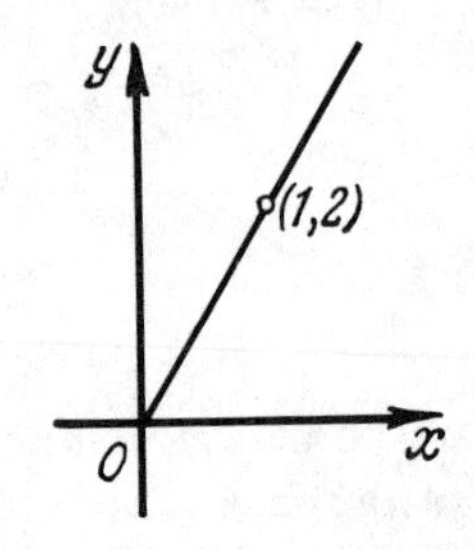

Fig. 5.9

EXAMPLE 5.3. Find the solution domain of the system

$$\begin{aligned} 2x - y &\geq 0 \\ -4x + 2y &\geq 0 \\ x + y &\geq 0. \end{aligned}$$

Proceeding as in example 5.2, we find that the solution domain is the single ray consisting of points of the form

$$B = t(1, 2) = (t, 2t) \quad (t \geq 0)$$

(fig. 5.9).

EXAMPLE 5.4. Find the solution domain of the system

$$\begin{aligned} 2x - y &\geq 0 \\ x + y &\geq 0 \\ -3x + y &\geq 0\,. \end{aligned}$$

In this case none of the equations

$$\begin{aligned} 2x - y &= 0\,, \\ x + y &= 0\,, \\ -3x + y &= 0\,, \end{aligned}$$

has a nontrivial solution which obeys all the given inequalities. Therefore, $\mathscr{K}_0$ consists of the single point (0, 0).

5.4. The Case Where the System (5.1) Is Not Normal

In this case, the solution domain $\mathscr{L}$ of the system (5.3) consists of more than just the origin. As a consequence, the equations (5.3) all define the same line in the plane, and this line is all of $\mathscr{L}$.

If $\mathscr{L}$ is not the y-axis, then none of the numbers $b_1, \ldots, b_m$ can be 0, and thus (5.3) reduces to

$$\begin{aligned} y &= -\frac{a_1}{b_1}x \\ y &= -\frac{a_2}{b_2}x \\ &\cdots\cdots\cdots \\ y &= -\frac{a_m}{b_m}x\,. \end{aligned} \tag{5.12}$$

Since each of these equations determines the same line,

$$-\frac{a_1}{b_1} = -\frac{a_2}{b_2} = \cdots = -\frac{a_m}{b_m} = r\,,$$

for some real number r. Thus the ith inequality of (5.1) is of one of the forms

$$y \geq rx - \frac{c_i}{b_i}, \qquad y \leq rx - \frac{c_i}{b_i},$$

depending on whether or not the sense of the inequality is reversed on division by b_i (that is, whether b_i is positive or negative).

Thus the lines bounding the half-planes determined by the inequalities of (5.1) are all parallel, so that $\mathscr{K}$ is either empty, a line, a half-plane, or a strip as in figure 5.10.

If $\mathscr{L}$ is the y-axis, then $b_1 = \cdots = b_m = 0$, so that the relevant boundaries are all parallel to the y-axis, and the above still holds.

According to lemma 5.1, for any point P in $\mathscr{K}$ the line $P + \mathscr{L}$ is contained in $\mathscr{K}$. Now consider any straight line $\mathscr{J}$ not parallel to $\mathscr{L}$.

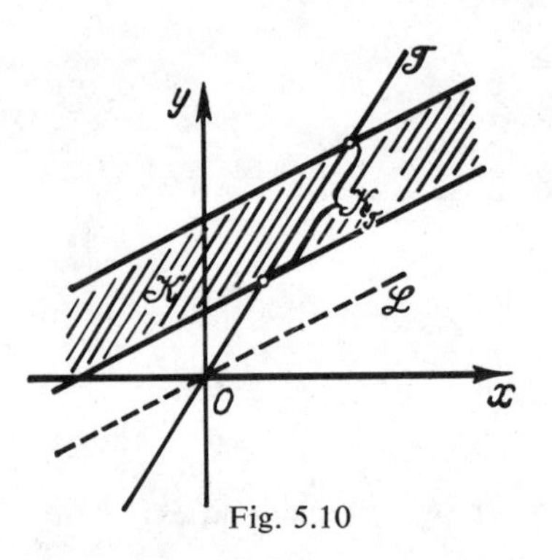

Fig. 5.10

If we knew which points of the line $\mathscr{J}$ belonged to $\mathscr{K}$ (we denote the set of these points by $\mathscr{K}_{\mathscr{J}}$), then we could determine $\mathscr{K}$ itself, because $\mathscr{K} = \mathscr{K}_{\mathscr{J}} + \mathscr{L}$ (fig. 5.10).

The equation of the line $\mathscr{L}$ is $a_1x + b_1y = 0$. In this equation, one of the coefficients, let us say b_1, is not zero. Thus, for the line $\mathscr{J}$ we may take the y-axis (with equation $x = 0$). In this case the set $\mathscr{K}_{\mathscr{J}}$, which we now denote by $\mathscr{K}_y$, is the part of the y-axis through $\mathscr{K}$. To find this set, we must set $x = 0$ in (5.1), yielding the inequalities

$$\begin{aligned} b_1y + c_1 &\geq 0 \\ b_2y + c_2 &\geq 0 \\ &\cdots\cdots\cdots \\ b_my + c_m &\geq 0 \end{aligned} \tag{5.13}$$

in the single unknown y, the solution of which presents no trouble. Note that the set $\mathscr{K}_y$ may be the empty set (in this case $\mathscr{K}$ is also empty), a point, a line segment, or a ray (but not the entire y-axis, for then $\mathscr{K}$ would be the entire plane, which is impossible). In determining this set, we determine the region $\mathscr{K}$ as well, since

$$\mathscr{K} = \mathscr{K}_y + \mathscr{L} \tag{5.14}$$

(if $\mathscr{L}$ is not parallel to the y-axis).

EXAMPLE 5.5. Find the solution domain $\mathscr{K}$ for the system

$$\begin{aligned} x + y - 1 &\geq 0 \\ -x - y + 2 &\geq 0 \\ 2x + 2y + 3 &\geq 0 . \end{aligned}$$

It is easily seen that the system is not normal and the set $\mathscr{L}$ is the line

$$x + y = 0$$

(which is not parallel to the y-axis). By setting $x = 0$, we obtain the system

$$\begin{aligned} y - 1 &\geq 0 \\ -y + 2 &\geq 0 \\ 2y + 3 &\geq 0 , \end{aligned}$$

from which it is clear that $\mathscr{K}_y$—the intersection of $\mathscr{K}$ with the y-axis—is a line segment with endpoints $C_1 = (0, 1)$ and $C_2 = (0, 2)$. This implies that $\mathscr{K}$ is the set of points of the form (fig. 5.11)

$$(0, y) + (x, -x) = (x, y - x) ,$$

where x is arbitrary and y is any number in the interval from 1 to 2 inclusive.

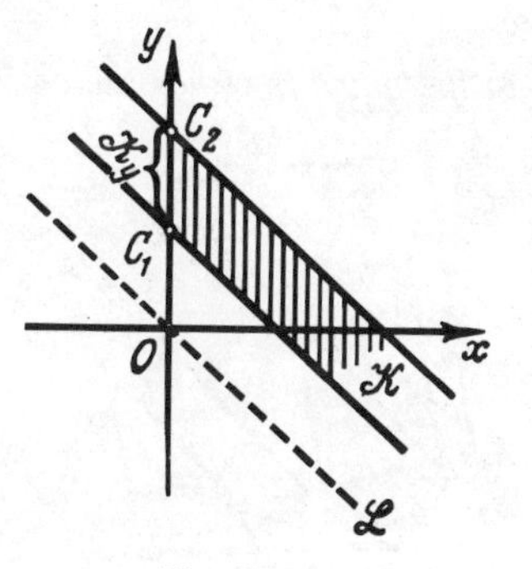

Fig. 5.11

In conclusion, we consider briefly a theorem which is a consequence of the above results. In the two-dimensional case we have considered, this theorem does not add much; however, it provides a convenient starting point for generalization to the n-dimensional case, which will be studied in more detail in chapter 7.

THEOREM 5.3. *Any (nonempty) convex polygonal region $\mathscr{K}$ in the plane can be represented as a sum*

$$\langle A_1, A_2, \ldots, A_p \rangle + (B_1, B_2, \ldots, B_q) . \tag{5.15}$$

(The first term of this sum is the convex hull of some point set $A_1, A_2, \ldots, A_p$, and the second is the set of points of the form $t_1B_1 + t_2B_2 + \cdots + t_qB_q$, where $t_1, t_2, \ldots, t_q$ are arbitrary nonnegative numbers.)

The proof of theorem 5.3 can be outlined in just a few words. Consider the system of linear inequalities determining $\mathscr{K}$. If this system is normal, then equation (5.7) is valid; by noting that in this formula $\mathscr{K}_0$ is a set of the form (B_1, B_2), (B_1), or (O) (the origin), we conclude that in the case of a normal system our assertion is correct. If the system is not normal, then we have equation (5.14), from which the representability of $\mathscr{K}$ in the desired form again follows. (Why?)

Note that if all the points $A_1, A_2, \ldots, A_p$ coincide with the origin, then the set $\langle A_1, A_2, \ldots, A_p \rangle$ also coincides with the origin; thus the sum (5.15) reduces to its second term. If the points $B_1, B_2, \ldots, B_q$ all coincide with the origin, then the set $(B_1, B_2, \ldots, B_q)$ also coincides with the origin and only the first term remains in equation (5.15).

The converse of the theorem is also true, although with one restriction.

THEOREM 5.4. *Any set of the form*

$$\langle A_1, A_2, \ldots, A_p \rangle + (B_1, B_2, \ldots, B_q)$$

in the plane is either the entire plane or a convex polygonal region.

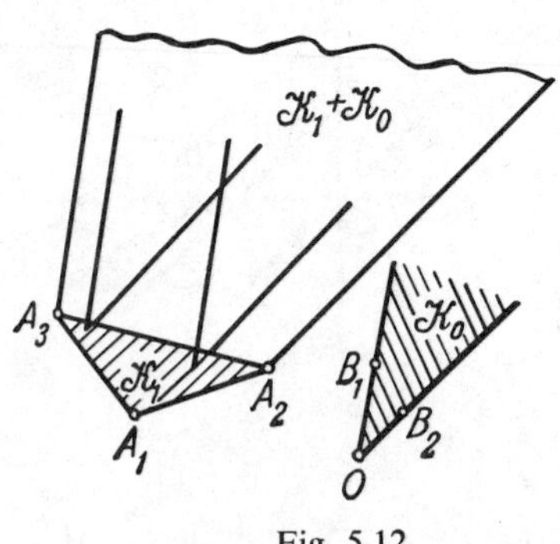

Fig. 5.12

The proof is quite clear. The second term, that is, the region $\mathscr{K}_0 = (B_1, B_2, \ldots, B_q)$, is the entire plane, a half-plane, an angular sector (smaller than 180°), a ray, or a point (the origin). The first term $\mathscr{K}_1 = \langle A_1, A_2, \ldots, A_p \rangle$ is a convex polygonal region. The set $\mathscr{K}_1 + \mathscr{K}_0$ can be obtained by translating $\mathscr{K}_0$ by the vectors $\boldsymbol{OK}_1$ (where K_1 is any point of $\mathscr{K}_1$) and combining the resulting sets (fig. 5.12). It is clear that this procedure leads either to the entire plane (if $\mathscr{K}_0$ is the entire plane), or to some convex polygonal region.

6 The Solution Domain of a System in Three Unknowns

Because the two-dimensional case has already been thoroughly treated, we are now in a position to treat the case of three unknowns with a minimum of effort.

Along with the given system

$$\begin{gathered} a_1 x + b_1 y + c_1 z + d_1 \geq 0 \\ \cdots\cdots\cdots\cdots\cdots\cdots\cdots\cdots \\ a_m x + b_m y + c_m z + d_m \geq 0\,, \end{gathered} \tag{6.1}$$

we also consider, as in chapter 5, the two systems

$$\begin{gathered} a_1 x + b_1 y + c_1 z \geq 0 \\ \cdots\cdots\cdots\cdots\cdots\cdots\cdots \\ a_m x + b_m y + c_m z \geq 0 \end{gathered} \tag{6.2}$$

and

$$\begin{gathered} a_1 x + b_1 y + c_1 z = 0 \\ \cdots\cdots\cdots\cdots\cdots\cdots\cdots \\ a_m x + b_m y + c_m z = 0\,. \end{gathered} \tag{6.3}$$

We again denote the solution domain of the system (6.1) by $\mathscr{K}$, that of (6.2) by $\mathscr{K}_0$, and that of (6.3) by $\mathscr{L}$. Making use of the terminology introduced earlier, we can say that $\mathscr{K}$ is a convex polyhedral region in three-space and that $\mathscr{K}_0$ is a convex polyhedral cone. As pointed out in chapter 5, lemmas 5.1 and 5.2 remain valid in this case, and thus so does theorem 5.1.

6.1. The Case Where the System (6.1) Is Normal

In this case, $\mathscr{K}_0$ contains no lines (by the analogue to theorem 5.1) and thus must have at least one vertex. Indeed, if $\mathscr{K}$ lies in a plane (the possibility of such a case was pointed out in chapter 2), then $\mathscr{K}$ is a convex polygonal region containing no line, and therefore, as in sec. 5.2, it must have a vertex. If $\mathscr{K}$ does not lie in a plane, we consider its boundary, which consists of polygonal plane sections, none of which contains a line. These polygonal plane sections, then, must have vertices, which must also be vertices of $\mathscr{K}$.

At least *three* plane faces must adjoin at any vertex A of $\mathscr{K}$; for if not, then all the plane faces passing through A would either coincide or intersect along a line. But then a small enough line segment passing through A and lying on the common face or on the common line would belong entirely to $\mathscr{K}$, contradicting the choice of A as a vertex.

All this makes it necessary for us to introduce some obvious changes into our procedure of sec. 5.2 for locating vertices. A *regular subsystem* must now consist of three (rather than of two) equations from the system

$$
\begin{aligned}
a_1x + b_1y + c_1z + d_1 &= 0 \\
\cdots\cdots\cdots\cdots&\cdots\cdots \\
a_mx + b_my + c_mz + d_m &= 0\,,
\end{aligned}
\tag{6.4}
$$

with the stipulation that the solution of this subsystem be unique. With this modification of the notion of a regular subsystem, our procedure remains essentially unchanged:

To find all vertices of the region $\mathscr{K}$, we solve all the regular subsystems of (6.4) *and discard those points which do not satisfy the original system* (6.1).

Theorem 5.2 remains valid; the changes which must be made in the proof are evident. The remark that a normal system has no solutions when $\mathscr{K}$ has no vertices is also correct in the present case.

EXAMPLE 6.1. Find the vertices of the region $\mathscr{K}$ determined by the system of inequalities

$$
\begin{aligned}
2x + y + z - 1 &\geq 0 \\
x + 2y + z - 1 &\geq 0 \\
x + y + 2z - 1 &\geq 0 \\
x + y + z - 1 &\geq 0\,.
\end{aligned}
\tag{6.5}
$$

The corresponding homogeneous equations are

$$\begin{aligned} 2x + y + z &= 0 \\ x + 2y + z &= 0 \\ x + y + 2z &= 0 \\ x + y + z &= 0\,. \end{aligned}$$

In solving this system, we find that the only solution is (0, 0, 0), so that (6.5) is normal.

In order to find the vertices we have to consider all possible three-equation subsystems of (6.4):

$$\begin{aligned} 2x + y + z - 1 &= 0 \qquad & 2x + y + z - 1 &= 0 \\ x + 2y + z - 1 &= 0 \qquad & x + 2y + z - 1 &= 0 \\ x + y + 2z - 1 &= 0\,; \qquad & x + y + z - 1 &= 0\,; \end{aligned}$$

$$\begin{aligned} 2x + y + z - 1 &= 0 \qquad & x + 2y + z - 1 &= 0 \\ x + y + 2z - 1 &= 0 \qquad & x + y + 2z - 1 &= 0 \\ x + y + z - 1 &= 0\,; \qquad & x + y + z - 1 &= 0\,. \end{aligned}$$

Carrying out the necessary calculations, we find that all the subsystems are regular, and that their solutions are

$$(\tfrac{1}{4}, \tfrac{1}{4}, \tfrac{1}{4}), \quad (0, 0, 1), \quad (0, 1, 0), \quad (1, 0, 0)\,,$$

of which only the last three satisfy (6.5). Consequently, the vertices of the region $\mathscr{K}$ are:

$$A_1 = (1, 0, 0), \qquad A_2 = (0, 1, 0), \qquad A_3 = (0, 0, 1)\,.$$

6.2. The Normal Homogeneous System of Inequalities (6.2)

Each of the inequalities in (6.2) determines a half-space whose boundary plane passes through the origin.

In this case, the intersection of the boundary planes must be a single point—the origin (since the system is normal). In other words, the set $\mathscr{K}_0$, which is the solution domain of (6.2), is a convex polyhedral cone with a *single* vertex. From the enumeration of the possible forms of convex polyhedral cones presented in chapter 4, it follows that $\mathscr{K}_0$ is an infinite convex pyramid, an angular sector in the plane, a ray, or a single point (the origin). Let us put aside the last case for the time being. In all the other cases we have

$$\mathscr{K}_0 = (B_1, B_2, \ldots, B_q),$$

where $B_1, B_2, \ldots, B_q$ is a set of points, one selected from each of the edges of the cone $\mathscr{K}_0$ (see theorem 4.2). We can find such points by proceeding as follows. Each (*a*) belongs to $\mathscr{K}_0$—that is, satisfies the system (6.2); and (*b*) belongs to the intersection of two distinct planes—that is, satisfies two independent[1] equations from the system (6.3).

If the only point satisfying conditions (*a*) and (*b*) is (0, 0, 0), then $\mathscr{K}_0$ reduces to the origin.

EXAMPLE 6.2. Find the solution domain $\mathscr{K}_0$ of the system

$$\begin{aligned} 2x + y + z &\geq 0 \\ x + 2y + z &\geq 0 \\ x + y + 2z &\geq 0 \\ x + y + z &\geq 0 \end{aligned} \tag{6.6}$$

and then the solution domain $\mathscr{K}$ of the system in example 6.1.

(Note that (6.6) is the homogeneous system corresponding to (6.5), and is thus normal.)

Pairs of independent equations can be selected in six different ways in this case:

$$\begin{aligned} x + 2y + z &= 0 \\ x + y + 2z &= 0; \end{aligned} \qquad \begin{aligned} 2x + y + z &= 0 \\ x + 2y + z &= 0; \end{aligned}$$

$$\begin{aligned} 2x + y + z &= 0 \\ x + y + 2z &= 0; \end{aligned} \qquad \begin{aligned} x + 2y + z &= 0 \\ x + y + z &= 0; \end{aligned}$$

$$\begin{aligned} 2x + y + z &= 0 \\ x + y + z &= 0; \end{aligned} \qquad \begin{aligned} x + y + 2z &= 0 \\ x + y + z &= 0. \end{aligned}$$

We select a nonzero solution (x, y, z) from each of these six subsystems and note that if (x, y, z) satisfies a given subsystem, then so does

1. The equations $ax + by + cz = 0$ and $a'x + b'y + c'z = 0$ are called "independent" if and only if at least one of the equalities in $a/a' = b/b' = c/c'$ is not valid; that is, if and only if they do not determine the same plane. Two such equations necessarily have a common solution (the origin), and thus their solution planes must intersect in a line.

$(-x, -y, -z)$. We thus have two distinct nonzero solutions to each subsystem. For example, in the first system we may take the points $(3, -1, -1)$ and $(-3, 1, 1)$; only the first of these satisfies (6.6). In this way we obtain the point $B_1 = (3, -1, -1)$. Treating the other five subsystems in the same way, we get the points $B_2 = (-1, 3, -1)$ and $B_3 = (-1, -1, 3)$. Hence the region $\mathscr{K}_0$ consists of all points of the form

$$t_1B_1 + t_2B_2 + t_3B_3 = (3t_1 - t_2 - t_3, -t_1 + 3t_2 - t_3, -t_1 - t_2 + 3t_3),$$

where t_1, t_2, and t_3 are any nonnegative numbers.

Let us turn now to the system of inequalities (6.5). As we have noted, the corresponding homogeneous system is precisely (6.6). As a consequence, the region $\mathscr{K}$ has the form

$$\langle A_1, A_2, A_3 \rangle + \mathscr{K}_0$$

and thus consists of the points

$$\begin{aligned} s_1A_1 + s_2A_2 + s_3A_3 + t_1B_1 + t_2B_2 + t_3B_3 \\ &= s_1(1, 0, 0) + s_2(0, 1, 0) + s_3(0, 0, 1) \\ &\quad + t_1(3, -1, -1) + t_2(-1, 3, -1) + t_3(-1, -1, 3) \\ &= (s_1 + 3t_1 - t_2 - t_3, s_2 - t_1 + 3t_2 - t_3, s_3 - t_1 - t_2 + 3t_3), \end{aligned}$$

where t_1, t_2, and t_3 are arbitrary nonnegative numbers and s_1, s_2, and s_3 are nonnegative numbers whose sum is 1.

6.3. The Case in Which the System of Inequalities (6.1) Is Not Normal

In this case the solution domain $\mathscr{L}$ of the homogeneous system of equations (6.3) contains some point other than the origin. Since $\mathscr{L}$ is the intersection of several planes passing through the origin, there are two possible cases:

1. $\mathscr{L}$ is a line passing through the origin. By the three-dimensional analogue to lemma 5.1, for each point P in $\mathscr{K}$, the straight line $P + \mathscr{L}$ is contained in $\mathscr{K}$. Consider a plane $\mathscr{J}$ which is not parallel to $\mathscr{L}$. If we knew which points of $\mathscr{J}$ belonged to the region $\mathscr{K}$ (we denote this set by $\mathscr{K}_{\mathscr{J}}$), then we could determine the region $\mathscr{K}$, for $\mathscr{K} = \mathscr{K}_{\mathscr{J}} + \mathscr{L}$ (fig. 6.1).

But any line is nonparallel to at least one of the coordinate (xy, xz, or yz) planes. Without loss of generality, let us suppose that $\mathscr{L}$ is not

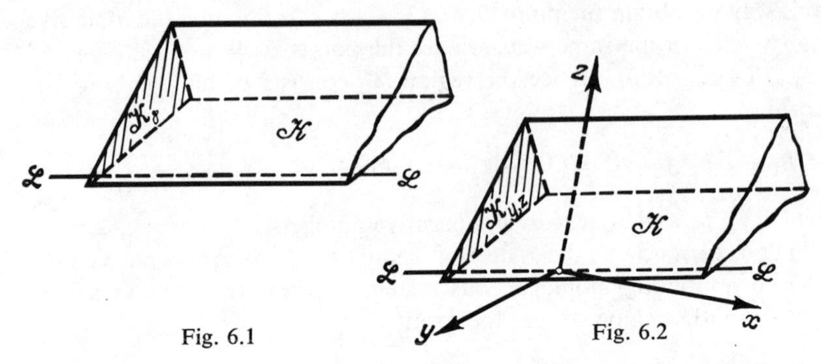

Fig. 6.1 Fig. 6.2

parallel to the yz-plane, which we select as the plane $\mathscr{J}$. Accordingly, we denote $\mathscr{K}_{\mathscr{J}}$ by $\mathscr{K}_{y,z}$, the portion of the yz-plane which is contained in $\mathscr{K}$ (fig. 6.2). To determine this set, we must set $x = 0$ in the system (6.1), yielding the system

$$\begin{aligned} b_1 y + c_1 z + d_1 &\geq 0 \\ \cdots\cdots&\cdots\cdots \\ b_m y + c_m z + d_m &\geq 0 \,, \end{aligned} \tag{6.7}$$

which we can solve by the method of chapter 5.

After determining the set $\mathscr{K}_{y,z}$, we may write

$$\mathscr{K} = \mathscr{K}_{y,z} + \mathscr{L}, \tag{6.8}$$

which describes $\mathscr{K}$ completely.

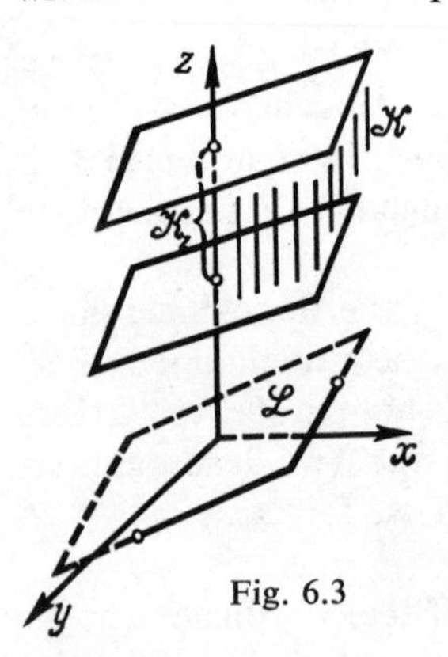

Fig. 6.3

2. $\mathscr{L}$ is a plane passing through the origin. Then all of the equations in (6.3) are equivalent and thus determine the same plane through the origin. Thus $\mathscr{K}$ is an intersection of half-spaces determined by parallel planes, and in analogy to the first case, these planes are nonparallel to at least one of the coordinate (x, y, or z) axes. Without loss of generality, we may assume that they are nonparallel to the z-axis. If $\mathscr{K}_z$ is the intersection of $\mathscr{K}$ with the z-axis (fig. 6.3), we may represent

$$\mathscr{K} = \mathscr{K}_z + \mathscr{L}.$$

We may determine $\mathscr{K}_z$ by setting $y = x = 0$ in (6.1):

$$\begin{aligned} c_1 z &\geq -d \\ &\cdots\cdots\cdot \\ c_m z &\geq -d_m \,. \end{aligned} \tag{6.9}$$

Solution of (6.9) thus determines $\mathscr{K}$.

REMARK 6.1. If the set $\mathscr{K}_{y,z}$ (or the set $\mathscr{K}_z$) is empty, then $\mathscr{K}$ is also empty. This means that the system (6.1) is inconsistent.

EXAMPLE 6.3. Find the solution domain $\mathscr{K}$ for the system

$$\begin{aligned} -2x + y + z - 1 &\geq 0 \\ -3x - y + 4z - 1 &\geq 0 \\ -x - 2y + 3z &\geq 0 \,. \end{aligned} \tag{6.10}$$

Consider the corresponding homogeneous equations

$$\begin{aligned} -2x + y + z &= 0 \\ -3x - y + 4z &= 0 \\ -x - 2y + 3z &= 0 \,. \end{aligned} \tag{6.11}$$

In solving this system, we note that the third equation is a consequence of the first two (subtract the first from the second), so that the system reduces to the first two equations. The solution domain $\mathscr{L}$ is thus the line of intersection of the two planes

$$\begin{aligned} -2x + y + z &= 0 \,, \\ -3x - y + 4z &= 0 \,. \end{aligned}$$

Take any point B on the line $\mathscr{L}$ which is distinct from the origin. (To do so, it is sufficient to find three numbers x, y, and z, not all zero, which satisfy the first two equations of (6.11).) For example, if we take the point (1, 1, 1), then $\mathscr{L}$ is the line OB, where $B = (1, 1, 1)$.

It is easily seen that the line $\mathscr{L}$ is not parallel to the yz-plane. Setting $x = 0$ in (6.10), we obtain the normal system

$$\begin{aligned} y + z - 1 &\geq 0 \\ -y + 4z - 1 &\geq 0 \\ -2y + 3z &\geq 0 \,, \end{aligned}$$

in the two unknowns y and z. Its solution domain $\mathscr{K}_{y,z}$ can be found by the method of chapter 5. By carrying out the necessary computations, we find that $\mathscr{K}_{y,z}$ consists of the single point $A = (0, \frac{3}{5}, \frac{2}{5})$. Hence the desired region, $\mathscr{K} = \mathscr{K}_{y,z} + \mathscr{L}$, consists of all points of the form

$$A + tB = (0, \tfrac{3}{5}, \tfrac{2}{5}) + t(1, 1, 1) = (t, \tfrac{3}{5} + t, \tfrac{2}{5} + t)\,,$$

where t is any real number ($\mathscr{K}$ is a line parallel to $\mathscr{L}$).

EXAMPLE 6.4. Find the solution domain $\mathscr{K}$ of the system

$$\begin{aligned} x - y + z + 1 &\geq 0\,, \\ -x + y - z + 2 &\geq 0\,. \end{aligned} \tag{6.12}$$

The corresponding homogeneous equations are

$$\begin{aligned} x - y + z &= 0\,, \\ -x + y - z &= 0\,. \end{aligned} \tag{6.13}$$

The second equation is a consequence of the first, so that the solution domain of (6.13) is the plane $\mathscr{L}$ determined by the equation

$$x - y + z = 0\,.$$

It is clear that this plane intersects the z-axis at a single point, so it cannot be parallel to this axis. To determine the set $\mathscr{K}_z$, we set $x = y = 0$ in (6.12), and we obtain

$$\begin{aligned} z + 1 &\geq 0\,, \\ -z + 2 &\geq 0\,, \end{aligned}$$

whence

$$-1 \leq z \leq 2\,. \tag{6.14}$$

Since $\mathscr{L}$, the solution domain of the first equation in (6.13), consists of all points of the form $(x, y, -x + y)$, $\mathscr{K}$ is the set $\mathscr{K}_z + \mathscr{L}$ consisting of all points of the form

$$(0, 0, z) + (x, y, -x + y) = (x, y, z - x + y)\,,$$

where x and y are arbitrary and z satisfies (6.14).

We shall conclude this section by stating two theorems generalizing the last two theorems of chapter 5 to the three-dimensional case. The only changes we must make are the replacement of the words "plane" and "polygonal" by the words "three-space" and "polyhedral."

THEOREM 6.1. *Any (nonempty) convex polyhedral region in three-space can be represented as a sum of the form*

$$\langle A_1, A_2, \ldots, A_p \rangle + (B_1, B_2, \ldots, B_q) .$$

THEOREM 6.2. *Any set of the form*

$$\langle A_1, A_2, \ldots, A_p \rangle + (B_1, B_2, \ldots, B_q)$$

in three-space is either all of space or a convex polyhedral region.

The proof of these theorems follows the proof for the two-dimensional case almost word for word; the reader should carry out the demonstration in detail.

7 Systems of Linear Inequalities in an Arbitrary Number of Unknowns

In the preceding section we have restricted our attention to systems of inequalities in two and three unknowns. This limitation was dictated by two circumstances: first of all, by the fact that investigation of these systems is simple and lies entirely within the framework of high school mathematics; and more importantly, by the fact that these systems have an intuitive geometric interpretation (in the plane or in three-space). In applications, however (for instance, to problems in linear programming), one often encounters systems of inequalities in $n > 3$ unknowns. To ignore these systems would greatly impoverish our understanding of the problem, and therefore we shall try, even if somewhat briefly, to describe the situation for an arbitrary n.

For the geometric interpretation of systems of inequalities in n unknowns, it is necessary to direct our attention to the so-called n-dimensional Euclidean space.

We begin by defining some of the more important concepts. A *point* in n-dimensional space is an ordered set of n numbers (n-tuple)

$$M = (x_1, x_2, x_3, \ldots, x_n) .$$

The number x_i is called the ith coordinate of M. The motivation for this definition lies in our description of points in the Euclidean plane as ordered pairs of real numbers, and of points in three-dimensional Euclidean space as ordered triples of real numbers. The point $(0, 0, \ldots, 0)$ is called the *origin of coordinates* or simply the *origin.*

We first define the concept of a *line segment* in n-dimensional space (or n-space). According to chapter 1, a line segment M_1M_2 in three-space can be characterized as the set of all points of the form

$$s_1M_1 + s_2M_2 ,$$

where s_1 and s_2 are nonnegative numbers adding up to 1. In going from three-space to n-space we adopt this characterization as the *definition* of a line segment. More precisely, let $M = (x_1, x_2, \ldots, x_n)$ and $M' = (x'_1, x'_2, \ldots, x'_n)$ be two arbitrary points in n-space. We define the line segment MM' as the set of all points of the form

$$sM + s'M' = (sx_1 + s'x'_1, sx_2 + s'x'_2, \ldots, sx_n + s'x'_n)\,, \quad (7.1)$$

where $s, s' \geq 0$ and $s + s' = 1$. For $s = 1$ and $s' = 0$, we get the point M; for $s = 0$ and $s' = 1$, we get the point M'. These two points are called the *endpoints* of the segment MM'; all remaining points (for which $s, s' > 0$) are called *interior points.*

For the definition of further notions relating to n-dimensional space, we require the concept of a *hyperplane*, a generalization of the notion of "plane" in three-dimensional space. The prefix *hyper* has a well-defined meaning; in n-dimensional space there can be "planes" of various kinds: one-dimensional planes ("lines"), two-dimensional planes, and so on. Finally, there are $(n - 1)$-dimensional planes, which we call hyperplanes.

DEFINITION 7.1. *A hyperplane in n-dimensional space is the set of all points* $M = (x_1, x_2, \ldots, x_n)$ *whose coordinates satisfy a linear equation*

$$a_1x_1 + a_2x_2 + \cdots + a_nx_n + b = 0\,, \quad (7.2)$$

in which at least one of the coefficients $a_1, a_2, \ldots, a_n$ *is different from zero.*

For $n = 3$ the equation (7.2) becomes $a_1x_1 + a_2x_2 + a_3x_3 + b = 0$, which is precisely the equation of a plane in three-space (here the coordinates are denoted by x_1, x_2, and x_3, rather than the usual x, y, and z).

The hyperplane (7.2) "separates" n-space into two parts: a region in which the inequality

$$a_1x_1 + a_2x_2 + \cdots + a_nx_n + b \geq 0 \quad (7.3)$$

is satisfied, and a region in which the opposite inequality

$$a_1x_1 + a_2x_2 + \cdots + a_nx_n + b \leq 0 \quad (7.4)$$

is satisfied. These regions are called *half-spaces*, and it is clear that their intersection is the determining hyperplane.

The concept of a convex region also carries over to the n-dimensional case. A set of points in n-space is called *convex* if for any two of its points M and M', it contains the entire segment MM'.

It is not hard to show that *every half-space is a convex set.* To do so, suppose that the points $M' = (x_1', x_2', \ldots, x_n')$ and $M'' = (x_1'', x_2'', \ldots, x_n'')$ belong to the half-space determined by equation (7.3). We shall show that every point M of the segment $M'M''$ also belongs to this half-space.

The coordinates of M can be represented in the form (7.1), or, equivalently, by

$$
\begin{aligned}
x_1 &= sx_1' + (1 - s)x_1'' , \\
x_2 &= sx_2' + (1 - s)x_2'' , \\
&\cdots\cdots\cdots\cdots\cdots\cdots \\
x_n &= sx_n' + (1 - s)x_n'' \qquad (0 \leq s \leq 1) .
\end{aligned}
$$

By inserting these expressions for the coordinates of M into the left-hand side of equation (7.3), we find that

$$
\begin{aligned}
&a_1x_1 + a_2x_2 + \cdots + a_nx_n + b \\
&\quad = a_1[sx_1' + (1 - s)x_1''] + a_2[sx_2' + (1 - s)x_2''] + \cdots \\
&\qquad + a_n[sx_n' + (1 - s)x_n''] + b \\
&\quad = s[a_1x_1' + a_2x_2' + \cdots + a_nx_n'] \\
&\qquad + (1 - s)[a_1x_1'' + a_2x_2'' + \cdots + a_nx_n''] + sb + (1 - s)b \\
&\quad = s[a_1x_1' + \cdots + a_nx_n' + b] + (1 - s)[a_1x_1'' + \cdots + a_nx_n'' + b] .
\end{aligned}
$$

Both of the sums in square brackets are nonnegative, since both M' and M'' belong to the half-space (7.3). Consequently, the entire expression is nonnegative (since $s \geq 0$ and $1 - s \geq 0$). This implies that all such points M belong to the half-space (7.3); that is, that this half-space is convex.

In view of this, it is not difficult to decide what geometric terminology should be applied to a system of linear inequalities in n unknowns:

$$
\begin{aligned}
a_1x_1 + a_2x_2 + \cdots + a_nx_n + a &\geq 0 , \\
b_1x_1 + b_2x_2 + \cdots + b_nx_n + b &\geq 0 , \\
\cdots\cdots\cdots\cdots\cdots\cdots\cdots\cdots &, \\
c_1x_1 + c_2x_2 + \cdots + c_nx_n + c &\geq 0 .
\end{aligned}
\qquad (7.5)
$$

Each of the above inequalities determines a half-space, and the solution domain $\mathscr{K}$ of the entire system of linear inequalities is an intersection of a finite number of half-spaces. The region $\mathscr{K}$ is convex, since all of the half-spaces entering into the intersection are convex.

In analogy with the three-dimensional case, we shall call a region in n-space which is the intersection of a finite number of half-spaces a *convex polyhedral region* and, when this intersection is bounded, a *convex polyhedron*. When we say that a region is "bounded," we shall mean that each coordinate of any point in the region is smaller in absolute value than some constant c: $|x_1| \leq c, \ldots, |x_n| \leq c$ for all points $(x_1, \ldots, x_n)$ of the region.

Thus, *the set $\mathscr{K}$ of all points in n-dimensional space whose coordinates satisfy* (7.5) *is the intersection of the half-spaces corresponding to the inequalities of the system, and therefore is a convex polyhedral region.*

The procedures for actually determining $\mathscr{K}$, which we considered in chapter 5 for the case of two unknowns and in chapter 6 for systems in three unknowns, carry over with only minor changes to the case of n unknowns. We shall not go into this in detail, however, for a complete treatment would require too many unnecessary pages of presentation. Moreover, when the number of unknowns becomes too large, these procedures become ineffective: their use requires an excessive amount of calculation.

It is worth noting that the general theorems regarding the structure of convex polyhedral regions in three-space carry over completely to n-space, although their proofs become more complicated. We shall limit ourselves to formulating these theorems and making some clarifying remarks.

THEOREM 7.1. *The convex hull of any set of points $A_1, A_2, \ldots, A_p$ is a convex polyhedron.*

We recall that the convex hull of the set of points $A_1, \ldots, A_p$ is the set of all points of the form

$$s_1A_1 + s_2A_2 + \cdots + s_pA_p\,,$$

where $s_1, s_2, \ldots, s_p$ are any nonnegative numbers whose sum is 1, and that this set is denoted by $\langle A_1, A_2, \ldots, A_p \rangle$. Theorem 7.1 establishes a connection between this notion and the differently defined notion of a convex polyhedron, the intersection of a finite number of half-spaces.

In two and three dimensions, the validity of theorem 7.1 is obvious (at least based on the intuitive picture of a convex hull); in the general case it is not at all obvious and requires proof.

THEOREM 7.2. (converse to theorem 7.1). *Every convex polyhedron is the convex hull of some subset of its points.*

In fact, we can make the stronger statement that *a convex polyhedron is the convex hull of its vertices.* The definition of a "vertex" carries over from that for the two-dimensional case: a *vertex* is a point of the polyhedron which is not an interior point of any segment entirely contained in the polyhedron. It can be shown that the number of vertices of a convex polyhedron is always finite.

THEOREM 7.3. *Every set of the form* $(B_1, B_2, \ldots, B_q)$ *is either all of n-space or a convex polyhedral cone with vertex at the origin.*

We recall that the symbol $(B_1, B_2, \ldots, B_q)$ denotes the set of all points of the form

$$t_1B_1 + t_2B_2 + \cdots + t_qB_q\,,$$

where $t_1, \ldots, t_q$ are any nonnegative numbers. A convex polyhedral cone is defined as the intersection of a finite number of half-spaces whose boundary hyperplanes have a point in common (the *vertex* of the cone). The validity of theorem 7.3 in three dimensions was established in chapter 4 (theorem 4.1).

THEOREM 7.4. *Any convex polyhedral cone with vertex at the origin can be represented as a set of the form* $(B_1, B_2, \ldots, B_q)$.

The validity of this assertion in the three-dimensional case was shown in chapter 4 (theorem 4.2).

THEOREM 7.5. *Any convex polyhedral region can be represented as a sum of the form*

$$\langle A_1, A_2, \ldots, A_p\rangle + (B_1, B_2, \ldots, B_q)\,.$$

THEOREM 7.6. *Every sum of the above form is either all of n-space or a convex polyhedral region.*

8 Inconsistent Systems

Up to this point, we have been interested primarily in systems of inequalities which have at least one solution (are consistent). The solution domains of such systems (in the plane and in three-space) are nonempty point sets. It would appear at first glance that the study of inconsistent systems is unnecessary, and it certainly seems unlikely that a very interesting theory could be built around them. But this is indeed only a superficial analysis of the situation. The properties of inconsistent systems not only are of interest for their own sake, but provide a key to the investigation of a large number of important assertions. The fundamental theorem of linear programming (the duality theorem of chapter 10), for example, ultimately reduces to the consideration of a special property of inconsistent systems.

Let us consider an arbitrary system of linear inequalities. For convenience of notation, we shall assume for the time being that the number of unknowns is three, although our discussion will apply just as well to systems in any number of unknowns.

Suppose, then, that the given system is

$$\begin{aligned} a_1x + b_1y + c_1z + d_1 &\geq 0 \\ a_2x + b_2y + c_2z + d_2 &\geq 0 \\ \cdots\cdots\cdots\cdots&\cdots\cdots \\ a_mx + b_my + c_mz + d_m &\geq 0 \,. \end{aligned} \tag{8.1}$$

Let us multiply the first of these inequalities by any nonnegative number k_1, the second by the nonnegative number k_2, and so on, finally adding all the inequalities thus obtained. The resultant inequality,

$$k_1(a_1x + b_1y + c_1z + d_1) + k_2(a_2x + b_2y + c_2z + d_2) + \cdots + k_m(a_mx + b_my + c_mz + d_m) \geq 0 \,, \tag{8.2}$$

or

$$(k_1a_1 + k_2a_2 + \cdots + k_ma_m)x + (k_1b_1 + k_2b_2 + \cdots + k_mb_m)y$$
$$+ (k_1c_1 + k_2c_2 + \cdots + k_mc_m)z + k_1d_1 + k_2d_2 + \cdots + k_md_m \geq 0 , \quad (8.3)$$

is called a *linear combination* of the inequalities in (8.1).

It may happen that some linear combination of the inequalities in (8.1) has the form

$$0x + 0y + 0z + d \geq 0 , \quad (8.4)$$

where d is strictly negative (after dividing by the strictly positive number $|d|$, we have the inequality $-1 \geq 0$). It is obvious that no choice of the numbers x, y, z can satisfy such an inequality, so that the system (8.1) is inconsistent. The remarkable fact is that the converse statement is also true: if the system (8.1) is inconsistent, then some linear combination of its members has the form (8.4).

We shall prove this statement in its general form (that is, for systems with any number of unknowns), but first we need a definition. We shall call the inequality

$$ax + by + cz + d \geq 0$$

contradictory if it is satisfied by no ordered triple (x, y, z) of real numbers. It is clear that any contradictory inequality has the form (8.4), where $d < 0$. (Prove this!) Our statement can now be reformulated as the following theorem.

THEOREM 8.1. *If a system of linear inequalities is inconsistent, then some linear combination of its members is contradictory.*

The proof is by induction on n, the number of unknowns.

For $n = 1$, such a system has the general form

$$\begin{aligned} a_1x + b_1 &\geq 0 \\ a_2x + b_2 &\geq 0 \\ &\cdots\cdots \\ a_mx + b_m &\geq 0 . \end{aligned} \quad (8.5)$$

We may assume that none of the coefficients $a_1, a_2, \ldots, a_m$ is zero. If, for example, $a_1 = 0$, then the first inequality has the form $0x + b_1 \geq 0$. If b_1 is nonnegative, then we can simply drop the inequality; if b_1 is negative, then the first inequality of our system (which is seen to be a linear combination of the members of (8.5) by setting $k_1 = 1, k_2 = k_3 = \cdots = k_m = 0$ in (8.2)) is contradictory and there is nothing to prove.

Assuming, then, that none of the coefficients $a_1, a_2, \ldots, a_m$ is zero, it is easily seen that among these coefficients some must be positive and some negative, for if all had the same sign, say positive, then (8.5) would reduce to the form

$$\begin{aligned} x &\geq -\frac{b_1}{a_1} \\ x &\geq -\frac{b_2}{a_2} \\ &\cdots\cdots\cdots \\ x &\geq -\frac{b_m}{a_m}. \end{aligned}$$

If, without loss of generality, we assume that $-(b_1/a_1)$ is the largest of the numbers $-(b_1/a_1), -(b_2/a_2), \ldots, -(b_m/a_m)$, the above system would be equivalent to the single noncontradictory inequality

$$x \geq -\frac{b_1}{a_1}.$$

As a consequence, (8.5) would necessarily be consistent.

By rearranging the inequalities, we may assume that the first k of the numbers $a_1, a_2, \ldots, a_m$ are positive and the remaining $m - k$ of them are negative. Then (8.5) is equivalent to the system

$$\begin{aligned} x &\geq -\frac{b_1}{a_1} \\ &\cdots\cdots\cdots \\ x &\geq -\frac{b_k}{a_k} \\ x &\leq -\frac{b_{k+1}}{a_{k+1}} \\ &\cdots\cdots\cdots\cdots \\ x &\leq -\frac{b_m}{a_m}. \end{aligned} \tag{8.6}$$

Among the numbers $-(b_1/a_1), \ldots, -(b_k/a_k)$ we select the largest, and we may rearrange the inequalities so that this number is $-(b_1/a_1)$.

Then the first k inequalities in (8.6) can be replaced by the first one. Similarly, we select the smallest of the numbers $-(b_{k+1}/a_{k+1}), \ldots, -(b_m/a_m)$, so that the remaining $m - k$ inequalities of (8.6) can be replaced by, for example, the last one. Then the original system (8.5) is equivalent to the following system of two inequalities:

$$x \geq -\frac{b_1}{a_1}$$

$$x \leq -\frac{b_m}{a_m},$$

or, in more compact form,

$$-\frac{b_1}{a_1} \leq x \leq -\frac{b_m}{a_m},$$

the inconsistency of which implies that

$$-\frac{b_1}{a_1} > -\frac{b_m}{a_m}, \tag{8.7}$$

From the inequality (8.7) it follows that

$$b_m a_1 - b_1 a_m < 0. \tag{8.8}$$

(Recall that $a_1 > 0$ and $a_m < 0$.) If we multiply the first inequality of (8.5) by the positive number $-a_m$ and the last by the positive number a_1, and then add them together, we obtain

$$0x + (b_m a_1 - b_1 a_m) \geq 0,$$

which, by virtue of the inequality (8.8), is contradictory. Hence the theorem is proved for systems of inequalities in one unknown.

It remains to show that if the theorem is valid for all systems of linear inequalities in $n - 1$ unknowns, then it is also true for all systems in n unknowns.

To carry out this inductive step, assume the theorem for systems in $n - 1$ unknowns and let an inconsistent system in n unknowns be given. Consider any one of its members, which must have the form

$$a_1 x_1 + \cdots + a_{n-1} x_{n-1} + a_n x_n + b \geq 0.$$

After transposing the term $a_n x_n$ to the right side, this becomes

$$a_1 x_1 + \cdots + a_{n-1} x_{n-1} + b \geq -a_n x_n .$$

If $a_n < 0$, we multiply both sides of the inequality by the positive number $-(1/a_n)$ to obtain an equivalent inequality of the form

$$a_1' x_1 + \cdots + a_{n-1}' x_{n-1} + b' \geq x_n .$$

If $a_n > 0$, we multiply both sides by the positive number $1/a_n$ to obtain an equivalent inequality of the form

$$-(a_1' x_1 + \cdots + a_{n-1}' x_{n-1} + b') \geq -x_n .$$

If $a_n = 0$, the inequality reads

$$a_1 x_1 + \cdots + a_{n-1} x_{n-1} + b \geq 0 ,$$

and we leave it as is. Thus, by multiplying each of the inequalities of the given system by some positive number (1 in the third case), we arrive at an equivalent system of the form

$$\begin{array}{c} P_1 \geq x_n \\ P_2 \geq x_n \\ \cdots\cdots\cdots \\ P_p \geq x_n \\ \\ -Q_1 \geq -x_n \\ -Q_2 \geq -x_n \\ \cdots\cdots\cdots \\ -Q_q \geq -x_n \\ \\ R_1 \geq 0 \\ R_2 \geq 0 \\ \cdots\cdots \\ R_r \geq 0 , \end{array} \tag{8.9}$$

where the P_α, the Q_β, and the R_γ are expressions of the form

$$a_1' x_1 + \cdots + a_{n-1}' x_{n-1} + b' .$$

By hypothesis, the original system is inconsistent. Thus, since each inequality in (8.9) is equivalent to its corresponding inequality in the original system, (8.9) must be inconsistent. It follows that the system

$$\begin{aligned}
&P_1 \geq Q_1, P_1 \geq Q_2, \ldots, P_1 \geq Q_q \\
&P_2 \geq Q_1, P_2 \geq Q_2, \ldots, P_2 \geq Q_q \\
&\cdots\cdots\cdots\cdots\cdots\cdots \\
&P_p \geq Q_1, P_p \geq Q_2, \ldots, P_p \geq Q_q \\
&\qquad R_1 \geq 0 \\
&\qquad \cdots\cdots \\
&\qquad R_r \geq 0
\end{aligned} \tag{8.10}$$

is inconsistent. For if (8.10) were to hold for some choice $x_1^0, \ldots, x_{n-1}^0$ of the $x_1, \ldots, x_{n-1}$, we would have the numerical inequalities

$$\begin{aligned}
&P_1^0 \geq Q_1^0, P_1^0 \geq Q_2^0, \ldots, P_1^0 \geq Q_q^0 \\
&P_2^0 \geq Q_1^0, P_2^0 \geq Q_2^0, \ldots, P_2^0 \geq Q_q^0 \\
&\cdots\cdots\cdots\cdots\cdots\cdots \\
&P_p^0 \geq Q_1^0, P_p^0 \geq Q_2^0, \ldots, P_p^0 \geq Q_q^0 \\
&\qquad R_1^0 \geq 0 \\
&\qquad \cdots\cdots \\
&\qquad R_r^0 \geq 0,
\end{aligned} \tag{8.11}$$

where the P_α^0, Q_β^0, and R_γ^0 are the values of P_α, Q_β, and R_γ, respectively, for the choices $x_1 = x_1^0, \ldots, x_{n-1} = x_{n-1}^0$. If we had the inequalities (8.11), then all the P_α^0 would lie to the right of all the Q_β^0 on the real number line (fig. 8.1), so that we could insert a real number x_n^0 "between" the greatest Q_β and the smallest P_α. Our previous choices of $x_1^0, \ldots, x_{n-1}^0$, along with this choice of x_n^0, would then give us a solution of the system

$$\begin{aligned}
&P_1 \geq x_n \geq Q_1, \ldots, P_1 \geq x_n \geq Q_q \\
&P_2 \geq x_n \geq Q_1, \ldots, P_2 \geq x_n \geq Q_q \\
&\cdots\cdots\cdots\cdots\cdots\cdots \\
&P_p \geq x_n \geq Q_1, \ldots, P_p \geq x_n \geq Q_q \\
&\qquad R_1 \geq 0 \\
&\qquad \cdots\cdots \\
&\qquad R_r \geq 0,
\end{aligned}$$

and thus of the inconsistent system (8.9), a contradiction.

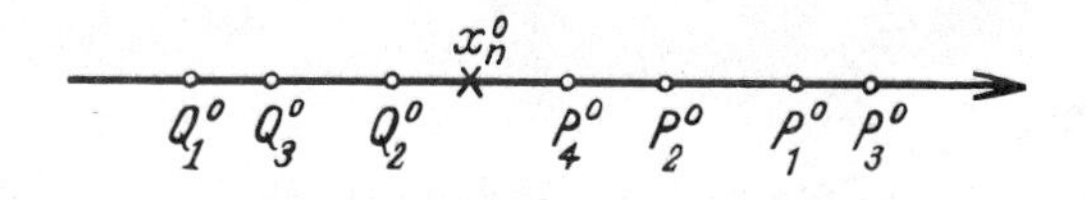

Fig. 8.1

Thus we have established that (8.10) is inconsistent. But since (8.10) is a system in $n - 1$ unknowns, we can apply the induction hypothesis to conclude that some linear combination of its inequalities is contradictory. Moreover, it is easily seen that each inequality in (8.10) is a linear combination of the inequalities in (8.9): to obtain $P_\alpha \geq Q_\beta$, we add the inequalities

$$P_\alpha \geq x_n,$$
$$-Q_\beta \geq -x_n;$$

we obtain $R_\gamma \geq 0$ directly from (8.9).

Consequently, some contradictory inequality is a linear combination of inequalities in (8.10), each of which is a linear combination of inequalities in (8.9), each of which is a linear combination of inequalities from the original system. It follows that some linear combination of inequalities from the original system is a contradictory inequality, and the theorem is proved.

This inconsistency theorem for systems of linear inequalities is only one of a number of close analogies between the properties of systems of linear inequalities and the properties of *systems of linear equations*. In the statement of the theorem, we may replace the word *inequality* by the word *equation* to obtain the following assertion:

THEOREM 8.2. *If a system of linear equations is inconsistent, then some linear combination of these equations is a contradictory equation.*

When worded somewhat differently, this assertion is known as the Kronecker-Capell theorem, a topic usually covered in courses on *linear algebra*. In order to arrive at a proper understanding of the above remarks, we must refine the concept of linear combination somewhat. Linear combinations of equations are formed in the same way as linear combinations of inequalities, except that we are permitted to multiply an equation by *any number whatsoever*, not merely by nonnegative numbers. As in the case of inequalities, an equation is called *contradictory* if it has no solutions. It is easy to show that a contradictory equation must be of the form

$$0x_1 + 0x_2 + \cdots + 0x_n + b = 0\,,$$

where b is nonzero (and by multiplying both sides by the nonzero number $1/b$ we obtain $1 = 0$).

One special case of the inconsistency theorem for inequalities is particularly important; this occurs when the system contains the inequalities

$$x_1 \geq 0, x_2 \geq 0, \ldots, x_n \geq 0\,. \tag{8.12}$$

By denoting the remaining part of the system by (S), we rephrase the problem as that of finding all *nonnegative* solutions of (S). If this system has no solutions, then by the theorem just proved, some linear combination of the inequalities of (S), say

$$a_1x_1 + a_2x_2 + \cdots + a_nx_n + a \geq 0\,, \tag{8.13}$$

added to some linear combination of the inequalities of (8.12), say

$$k_1x_1 + k_2x_2 + \cdots + k_nx_n \geq 0\,,$$

where $k_1, k_2, \ldots, k_n$ are nonnegative, gives the contradictory inequality

$$0x_1 + 0x_2 + \cdots + 0x_n + c \geq 0\,,$$

where c is a negative number. Consequently,

$$a_1 = -k_1 \leq 0, a_2 = -k_2 \leq 0, \ldots, a_n = -k_n \leq 0, a < 0\,.$$

We shall express this result in the following form:

COROLLARY 8.1 (to theorem 8.1). *If a system of inequalities has no nonnegative solutions, then some linear combinations of these inequalities has the form* (8.13), *where all the coefficients* $a_1, a_2, \ldots, a_n \leq 0$, *and the free term* $a < 0$.

For if the given system has no nonnegative solutions, introduction of the inequalities in (8.12) yields an inconsistent system, and the above argument applies.

Another important consequence of the theorem is the connection that can be established between a given system of inequalities and another system which includes equations as well. We illustrate this by means of an example of the form (8.1) (in three unknowns x, y, and z).

If (8.1) is inconsistent, then some linear combination (8.3) of its inequalities has the form (8.4), where $d < 0$. This means that there exist nonnegative numbers $k_1, k_2, \ldots, k_m$ for which

$$\begin{aligned} k_1a_1 + k_2a_2 + \cdots + k_ma_m &= 0\,, \\ k_1b_1 + k_2b_2 + \cdots + k_mb_m &= 0\,, \\ k_1c_1 + k_2c_2 + \cdots + k_mc_m &= 0\,, \\ d = k_1d_1 + k_2d_2 + \cdots + k_md_m &< 0\,. \end{aligned}$$

In other words, the mixed system

$$\begin{aligned} a_1y_1 + a_2y_2 + \cdots + a_my_m &= 0 \\ b_1y_1 + b_2y_2 + \cdots + b_my_m &= 0 \\ c_1y_1 + c_2y_2 + \cdots + c_my_m &= 0 \\ d_1y_1 + d_2y_2 + \cdots + d_my_m &= d \\ y_1 \qquad\qquad\qquad &\geq 0 \\ y_2 \qquad\qquad &\geq 0 \\ \cdots\cdots\cdots\cdots\cdots\cdots & \\ y_m &\geq 0\,, \end{aligned} \tag{8.14}$$

consisting of equations and inequalities has the solution $y_1 = k_1, y_2 = k_2, \ldots, y_m = k_m$. Multiplying this solution by the positive number $-(1/d)$, we obtain a solution of the system

$$\begin{aligned} a_1y_1 + a_2y_2 + \cdots + a_my_m &= 0 \\ b_1y_1 + b_2y_2 + \cdots + b_my_m &= 0 \\ c_1y_1 + c_2y_2 + \cdots + c_my_m &= 0 \\ d_1y_1 + d_2y_2 + \cdots + d_my_m &= -1 \\ y_1 \qquad\qquad\qquad &\geq 0 \\ y_2 \qquad\qquad &\geq 0 \\ \cdots\cdots\cdots\cdots\cdots\cdots & \\ y_m &\geq 0\,, \end{aligned} \tag{8.15}$$

which must, therefore, be consistent.

Consequently, *if* (8.1) *is inconsistent, then* (8.15) *is consistent.* The converse is also true: if (8.15) is consistent, then the system

$$\begin{aligned} a_1 y_1 + a_2 y_2 + \cdots + a_m y_m &= 0 \\ b_1 y_1 + b_2 y_2 + \cdots + b_m y_m &= 0 \\ c_1 y_1 + c_2 y_2 + \cdots + c_m y_m &= 0 \\ d_1 y_1 + d_2 y_2 + \cdots + d_m y_m &= -1 \end{aligned} \tag{8.16}$$

of linear equations has a nonnegative solution, say $y_1 = k_1, \ldots, y_m = k_m$; then the linear combination (8.3) of the inequalities in (8.1) reduces to the contradictory inequality

$$0x + 0y + 0z - 1 \geq 0 ,$$

so that (8.1) is inconsistent.

9 Dual Convex Polyhedral Cones

In chapter 4 we promised to discuss convex polyhedral cones in more detail; we shall do so here.

Every convex polyhedral cone with vertex at the origin in three-space, as has been noted earlier, is the solution domain of some system of linear inequalities of the form

$$\begin{aligned} a_1x + b_1y + c_1z &\geq 0 \\ a_2x + b_2y + c_2z &\geq 0 \\ \cdots\cdots&\cdots\cdots \\ a_mx + b_my + c_mz &\geq 0\,. \end{aligned} \tag{9.1}$$

Along with this system, consider some given inequality

$$ax + by + cz \geq 0\,. \tag{9.2}$$

We shall say that the inequality (9.2) *is a consequence of* (9.1) *if every set of values x, y, z satisfying* (9.1) *satisfies* (9.2) *as well.*

Of course, every linear combination of the inequalities of (9.1) is a consequence of (9.1), but is the converse true? It appears that the answer is yes.

THEOREM 9.1. *Any homogeneous inequality* (9.2) *which is a consequence of a system of homogeneous inequalities* (9.1) *can be expressed as a linear combination of these inequalities.*

Proof. For ease of notation, let us denote the left sides of the first, second, . . ., mth inequality of (9.1) by $P_1, P_2, \ldots, P_m$, respectively, and the left side of (9.2) by P. Then (9.1) becomes

$$\begin{aligned} P_1 &\geq 0 \\ P_2 &\geq 0 \\ \cdots&\cdots \\ P_m &\geq 0\,, \end{aligned}$$

and (9.2) becomes

$$P \geq 0 .$$

Given that $P \geq 0$ is a consequence of (9.1), we know that each solution of (9.1) satisfies $P \geq 0$ and thus cannot satisfy $P = -1$. In other words, the mixed system

$$\begin{aligned} P_1 &\geq 0 \\ P_2 &\geq 0 \\ &\cdots\cdots \\ P_m &\geq 0 \\ P &= -1 , \end{aligned} \tag{9.3}$$

is inconsistent. Our tactic will be to apply our theorem on inconsistent systems to the system (9.3). Of course, this cannot be done immediately, for the theorem refers to systems consisting of *inequalities* only, whereas (9.3) contains the equation $P = -1$. However, this equation is equivalent to the system of inequalities

$$\begin{aligned} P &\geq -1 \\ P &\leq -1 \end{aligned} \quad \text{or} \quad \begin{aligned} P + 1 &\geq 0 \\ -P - 1 &\geq 0 , \end{aligned}$$

so that (9.3) is equivalent to the system

$$\begin{aligned} P_1 &\geq 0 \\ P_2 &\geq 0 \\ &\cdots\cdots \\ P_m &\geq 0 \\ P + 1 &\geq 0 \\ -P - 1 &\geq 0 , \end{aligned} \tag{9.4}$$

which is inconsistent because (9.3) is.

According to the inconsistency theorem, some linear combination of the inequalities in (9.4) is contradictory. In other words, there are non-negative numbers $k_1, k_2, \ldots, k_m, k_{m+1}, k_{m+2}$ such that the inequality

$$k_1P_1 + k_2P_2 + \cdots + k_mP_m + k_{m+1}(P + 1) + k_{m+2}(-P - 1) \geq 0 \tag{9.5}$$

reduces to

$$0x + 0y + 0z + d \geq 0 ,$$

where the number d is negative. Multiplying $k_1, \ldots, k_m, k_{m+1}, k_{m+2}$ by the positive number $-(1/d)$ if necessary, we may assume that (9.5) reduces to

$$0x + 0y + 0z - 1 \geq 0\,,$$

so that

$$k_1P_1 + k_2P_2 + \cdots + k_mP_m + (k_{m+1} - k_{m+2})P = 0x + 0y + 0z\,,$$

and the constant term $k_{m+1} - k_{m+2} = -1$. This yields

$$P = k_1P_1 + \cdots + k_mP_m\,,$$

expressing (9.2) as a linear combination (recall that the k_i are nonnegative) of the inequalities in (9.1).

The theorem we have just proved is interesting in itself, but its geometric content is even more fascinating. In order to investigate this aspect, we must first digress briefly into a discussion of analytic geometry. Let

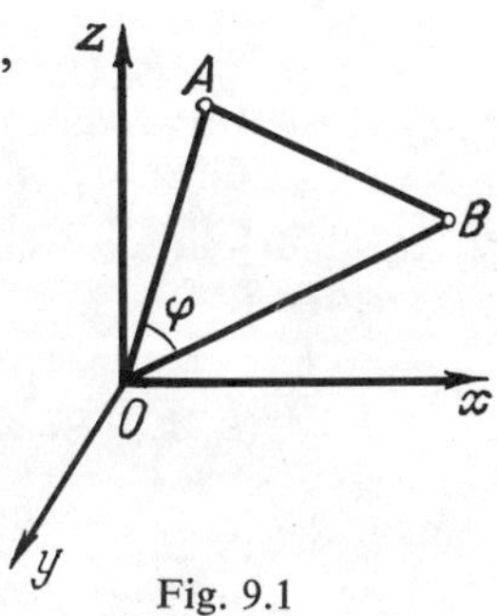

Fig. 9.1

$$A = (x_A, y_A, z_A), \quad B = (x_B, y_B, z_B)$$

be two distinct points in space, both different from the origin O.

We now apply the "law of cosines" to the triangle OAB. If φ is the angle between the segments OA and OB, then

$$\|AB\|^2 = \|OA\|^2 + \|OB\|^2 - 2\|OA\|\,\|OB\| \cos\varphi\,. \tag{9.6}$$

But

$$\|OA\|^2 = x_A^2 + y_A^2 + z_A^2\,;$$
$$\|OB\|^2 = x_B^2 + y_B^2 + z_B^2\,;$$
$$\|AB\|^2 = (x_A - x_B)^2 + (y_A - y_B)^2 + (z_A - z_B)^2\,,$$

so that, after expanding and combining similar terms in (9.6), we obtain

$$-2(x_Ax_B + y_Ay_B + z_Az_B) = -2\|OA\|\,\|OB\| \cos\varphi\,. \tag{9.7}$$

The angle φ is right or obtuse if and only if $\cos\varphi \leq 0$, so it follows from (9.7) that *the angle between the segments OA and OB is right or* *obtuse if and only if*

$$x_A x_B + y_A y_B + z_A z_B \leq 0 .$$

We shall, in what follows, denote the expression on the left hand side of this inequality simply by (A, B):

$$(A, B) = x_A x_B + y_A y_B + z_A z_B .$$

Thus, the above statement can be rephrased as follows:

The angle between the segments OA and OB is right or obtuse if and only if

$$(A, B) \leq 0 .$$

In what follows, we shall need one further property of the quantity (A, B), that

$$(k_1 A_1 + k_2 A_2, B) = k_1(A_1, B) + k_2(A_2, B) , \tag{9.8}$$

for any real numbers k_1 and k_2. The proof is almost obvious: Since the point $k_1 A_1 + k_2 A_2$ has coordinates

$$(k_1 x_{A_1} + k_2 x_{A_2}, k_1 y_{A_1} + k_2 y_{A_2}, k_1 z_{A_1} + k_2 z_{A_2}) ,$$

we have

$$\begin{aligned}
&(k_1 A_1 + k_2 A_2, B) \\
&\quad = (k_1 x_{A_1} + k_2 x_{A_2}) x_B + (k_1 y_{A_1} + k_2 y_{A_2}) y_B + (k_1 z_{A_1} + k_2 z_{A_2}) z_B \\
&\quad = k_1(x_{A_1} x_B + y_{A_1} y_B + z_{A_1} z_B) + k_2(x_{A_2} x_B + y_{A_2} y_B + z_{A_2} z_B) \\
&\quad = k_1(A_1, B) + k_2(A_2, B) .
\end{aligned}$$

We now turn to the main topic of this section—convex polyhedral cones in three-space. According to definition 4.1, a convex polyhedral cone with vertex at the point S is an intersection of a finite number of half-spaces whose boundary planes all pass through S. The "most typical" example of such an object, as we know, is an infinite convex pyramid. Throughout this section it will be assumed that the vertex of any convex polyhedral cone considered lies at the origin O.

Let $\mathscr{K}$ be such a cone, and pick some point B, distinct from the origin, with the property that *the angle between the segments OB and OA is right or obtuse for each point A of $\mathscr{K}$.* Such a point B can always be found. If we pass a plane π through the vertex O so that $\mathscr{K}$ lies entirely in one of the two resultant half-spaces (fig. 9.2), then the *normal* (perpendicular) ray to the plane π extending into the other half-space consists entirely of points B of the desired type.

We now consider the collection of all points B which possess the above property supplemented by one additional point—the origin—and denote the resulting object by $\mathscr{K}^*$. We shall first prove the following

LEMMA 9.1. *$\mathscr{K}^*$ is itself a convex polyhedral cone* (see fig. 9.3).

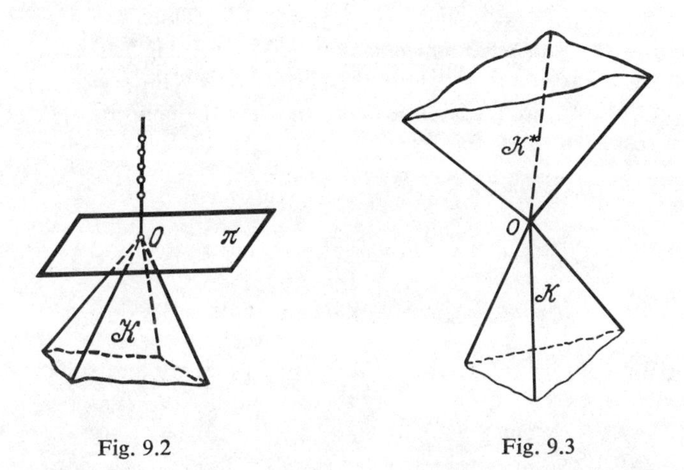

Fig. 9.2 Fig. 9.3

Proof. By theorem 4.2, every convex polyhedral cone $\mathscr{K}$ is a set of the form $(A_1, A_2, \ldots, A_m)$. This means that any point $A \in \mathscr{K}$ can be represented as

$$A = t_1A_1 + t_2A_2 + \cdots + t_mA_m , \tag{9.9}$$

where $t_1, t_2, \ldots, t_m$ are nonnegative real numbers. If the point B belongs to the set $\mathscr{K}^*$, then the angle between the segment OB and each of the segments OA for $A \in \mathscr{K}$ is right or obtuse; that is,

$$(A, B) \leq 0 \qquad \text{for all } A \in \mathscr{K}.$$

Since the points A of $\mathscr{K}$ are exactly those points which can be represented in the form (9.9), and since for such a point A,

$$(A, B) = t_1(A_1, B) + t_2(A_2, B) + \cdots + t_m(A_m, B)$$

by (9.8), we have

$$t_1(A_1, B) + t_2(A_2, B) + \cdots + t_m(A_m, B) \leq 0\,, \tag{9.10}$$

for all nonnegative choices of $t_1, t_2, \ldots, t_m$. In particular, we may successively "single out" the points $A_1, A_2, \ldots, A_m$ (by setting the appropriate coefficient equal to 1 and the rest equal to 0) to conclude that

$$(A_1, B) \leq 0, (A_2, B) \leq 0, \ldots, (A_m, B) \leq 0\,. \tag{9.11}$$

But conversely, if (9.11) is valid, then so is (9.10) (for any nonnegative numbers $t_1, t_2, \ldots, t_m$); that is to say, $B \in \mathscr{K}^*$. Thus, *for a point B to belong to $\mathscr{K}^*$, it is necessary and sufficient that* (9.11) *be valid.*

Let the coordinates of the point A_i be (a_i, b_i, c_i) (for $i = 1, 2, \ldots, m$), and let the coordinates of B be (x, y, z). Then the conditions (9.11) can be expressed in the form

$$\begin{aligned} a_1x + b_1y + c_1z &\leq 0\,, \\ a_2x + b_2y + c_2z &\leq 0\,, \\ &\cdots\cdots\cdots \\ a_mx + b_my + c_mz &\leq 0\,. \end{aligned} \tag{9.12}$$

Thus, for B to belong to the set $\mathscr{K}^*$, it is necessary and sufficient that its coordinates x, y, and z satisfy the inequalities (9.12):

$\mathscr{K}^*$ is the solution domain of the system (9.12).

Since (9.12) is homogeneous, its solution domain is a convex polyhedral cone in three-space, so that $\mathscr{K}^*$ is a convex polyhedral cone, which is what we set out to prove.

Therefore, *to each convex polyhedral cone $\mathscr{K}$ we can associate another such cone $\mathscr{K}^*$, consisting of all points B for which the segment OB makes a right or obtuse angle with any segment OA, where $A \in \mathscr{K}$.*

We shall say that the cone $\mathscr{K}^*$ is *conjugate* to $\mathscr{K}$.

One question naturally arises: what is the cone conjugate to $\mathscr{K}^*$? That is, what can one say about $(\mathscr{K}^*)^*$? From the definition of $\mathscr{K}^*$, it immediately follows that the set $(\mathscr{K}^*)^*$ must contain the initial point set $\mathscr{K}$. (Why?) But the assertion that these sets will exactly coincide is not immediately obvious. Moreover, it is not at all easy to give a geometric argument for this assertion. In any case, we choose to prove the equality of $(\mathscr{K}^*)^*$ and $\mathscr{K}$ algebraically on the basis of theorem 9.1; actually, as will be clear from the proof below, *the geometric content of theorem 9.1 for systems in three unknowns consists precisely of the statement $(\mathscr{K}^*)^* = \mathscr{K}$.*

THEOREM 9.2. *Let $\mathcal{K}$ be a convex polyhedral cone. Then the sets $(\mathcal{K}^*)^*$ and $\mathcal{K}$ are identical.*

Proof of theorem 9.2. Let $C = (a, b, c)$ be any point of the set $(\mathcal{K}^*)^*$. Then each point $B = (x, y, z)$ of $\mathcal{K}^*$ must satisfy the relation $(B, C) \leq 0$; that is,

$$ax + by + cz \leq 0 . \tag{9.13}$$

Since $\mathcal{K}^*$ is the solution domain of (9.12), and since every $B \in \mathcal{K}^*$ satisfies (9.13), every solution of (9.12) must also satisfy (9.13). In other words, (9.13) *is a consequence* of (9.12).

But by theorem 9.1, this is only possible when (9.13) is a linear combination of the inequalities in (9.12); that is, when

$$(a, b, c) = t_1(a_1, b_1, c_1) + t_2(a_2, b_2, c_2) + \cdots + t_m(a_m, b_m, c_m) ,$$

where $t_1, t_2, \ldots, t_m$ are some nonnegative real numbers. But the last equation implies that

$$C = t_1A_1 + t_2A_2 + \cdots + t_mA_m ;$$

that is to say, that the point C belongs to the cone $\mathcal{K}$. Hence, any point C which belongs to $(\mathcal{K}^*)^*$ also belongs to $\mathcal{K}$. The converse, that $\mathcal{K}$ is contained in $(\mathcal{K}^*)^*$, was shown earlier. Therefore $\mathcal{K} = (\mathcal{K}^*)^*$, and the theorem is proved.

10 The Duality Theorem of Linear Programming

Linear programming is a relatively new area of applied mathematics; it has been developed over the last fifteen to twenty years in connection with the solution of various problems in economics.

As a rule, problems which one meets in economics (and especially in industrial planning) involve the calculation of *extrema* (maxima and minima), and are usually directed toward finding the most *profitable* alternatives. For example, simplifying (and even exaggerating) the actual situation, we may suppose that in some business which manufactures items of two kinds, the daily productive capacity of a factory is limited to 100 items of the first sort or 300 of the second. Meanwhile, the quality control division cannot check more than 150 items (of either kind) per day. One knows, in addition, that the items of the first kind cost twice as much as those of the second kind. Under these conditions, we wish to determine which production plan will secure the business the most profit; that is, how many items of the first type and how many of the second should be produced each day if our goal is to maximize profits? Until recently, the only means for solving such problems was by trial and error, but the situation has changed. In recent decades, the complexity of industrial activities has increased to such an extent that simple trial and error procedures are impossible. There are so many factors influencing the outcome of a given course of action that the number of trials would have to reach into the millions. These problems have greatly stimulated interest in mathematical methods in economics.

Let us return to the example given above. The desired plan of production can be described by two nonnegative integers x, y (x is the number of items of the first kind, y is the number of the second), which must satisfy the following conditions

1) $3x + y \leq 300$;[1]

2) $x + y \leq 150$;

3) $2x + y$ is at a maximum.

In other words, from among the nonnegative integral solutions of the system

$$\begin{aligned} 3x + y &\leq 300 \\ x + y &\leq 150\,, \end{aligned} \tag{10.1}$$

we must select the one that gives the function

$$f(x, y) = 2x + y$$

as large a value as possible.

The solution domain of the system (10.1) in the first quadrant (the set of nonnegative solutions of (10.1)) can be represented as the shaded polygon in figure 10.1. From this sketch it is clear that the solution to the problem is the point $P = (75, 75)$—one of the vertices of the polygon.

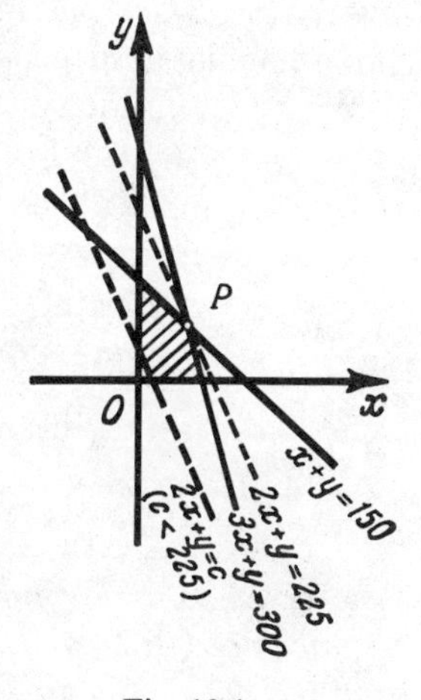

Fig. 10.1

Indeed, consider the straight line l_c defined by the equation $2x + y = c$, where c is some constant. As the number c increases, the line l_c moves upward (remaining parallel to its original position). The largest value of c for which the line l_c still has a point in common with the shaded polygon is that value for which l_c passes through the point P. Consequently, the function $2x + y$ attains its largest value at this point (in comparison with its values elsewhere in the polygon).

The example we have chosen is of course very primitive, but never-

1. The conditions originate from the assembly line. Recall that if the shop were to limit itself to only one item, it could produce either 100 items of the first kind or 300 of the second kind. If production is mixed, then one item of the first kind is interchangeable with three of the second in terms of assembly-line time. Consequently, the entire output of the shop in units of the second type of item consists of $3x + y$ items; this number cannot exceed 300.

theless it provides some idea of the character of the problems dealt with in linear programming. In such problems, one is required to find the maximum or minimum value of some linear function in n unknowns:

$$f(x_1, x_2, \ldots, x_n) = c_1x_1 + c_2x_2 + \cdots + c_nx_n\,,$$

under the condition that these variables satisfy some system of linear inequalities (which includes the conditions that the variables are nonnegative: $x_1 \geq 0, x_2 \geq 0, \ldots, x_n \geq 0$).

The methods for solving problems in linear programming that have been developed up to the present time are quite reliable. The majority of them are quite elementary and fit easily into the framework of high school mathematics.

The purposes of this book do not include a basic exposition of the theory of linear programming; many books and monographs, including a number of popular ones, have been devoted to this topic. Here we shall touch only on a single fragment of this theory—the so-called *duality principle*. But first, we shall formulate the *problem* of linear programming in its general form.

We are given the system

$$\begin{gathered} a_{11}x_1 + a_{12}x_2 + \cdots + a_{1n}x_n + b_1 \geq 0 \\ a_{21}x_1 + a_{22}x_2 + \cdots + a_{2n}x_n + b_2 \geq 0 \\ \cdots\cdots\cdots\cdots\cdots\cdots\cdots\cdots \\ a_{m1}x_1 + a_{m2}x_2 + \cdots + a_{mn}x_n + b_m \geq 0 \end{gathered} \tag{10.2}$$

of m linear inequalities in n unknowns,[2] and some linear function

$$f(x_1, x_2, \ldots, x_n) = c_1x_1 + c_2x_2 + \cdots + c_nx_n\,.$$

The problem is *to find all nonnegative* ($x_1 \geq 0$, $x_2 \geq 0, \ldots, x_n \geq 0$) *solutions of* (10.2) *which give the function f its greatest possible value; that is, which maximize f.*

We call this the *original* problem or *problem A*.

We now associate with problem A another problem A', its *dual problem*, which is formulated as follows. Given the system

$$\begin{gathered} a_{11}y_1 + a_{21}y_2 + \cdots + a_{m1}y_m + c_1 \leq 0 \\ a_{12}y_1 + a_{22}y_2 + \cdots + a_{m2}y_m + c_2 \leq 0 \\ \cdots\cdots\cdots\cdots\cdots\cdots\cdots\cdots \\ a_{1n}y_1 + a_{2n}y_2 + \cdots + a_{mn}y_m + c_n \leq 0 \end{gathered} \tag{10.3}$$

2. The coefficient of the jth unknown in the ith inequality of the system (10.2) is denoted by a_{ij}.

of n linear inequalities in m unknowns, and the linear function

$$g(y_1, y_2, \ldots, y_m) = b_1 y_1 + b_2 y_2 + \cdots + b_m y_m,$$

find all nonnegative solutions of (10.3) *which give the function its smallest possible value, that is, which minimize g.*

In comparing problems A and A', we note the following:

1. The coefficient of the jth unknown in the ith equation of (10.2) is the same as the coefficient of the ith unknown in the jth equation of (10.3).

2. The free terms in the inequalities in each of these problems are the coefficients of the unknowns in the linear function of the other problem.

3. The inequalities of problem A are all of the type ≥ 0, and in this problem we attempt to maximize f. The inequalities of problem A', on the other hand, are all of the form ≤ 0, and we are required to minimize g.

The *duality theorem* is stated as follows:

If the original problem has a solution, then the dual problem also has a solution. Moreover, the maximum value of the function f is the minimum value of the function g; that is,

$$\max f = \min g.$$

We shall prove this theorem by reducing it to the question of the consistency of a certain system of inequalities.

To make the organization of the proof more obvious, we shall break it down into several steps.

Step 1: LEMMA. *If* $(x_1^0, \ldots, x_n^0)$ *is any nonnegative solution of* (10.2), *and* $(y_1^0, \ldots, y_m^0)$ *is any nonnegative solution of* (10.3), *then the values of f and g for these solutions are related by the inequality*

$$f_0 \leq g_0,$$

where $f_0 = f(x_1^0, \ldots, x_n^0)$ *and* $g_0 = g(y_1^0, \ldots, y_m^0)$.

Proof. Consider the inequalities of system (10.2), substituting $x_1^0, \ldots, x_n^0$ for $x_1, \ldots, x_n$, respectively. Multiply the first inequality by y_1^0, the second by y_2^0, etc., and then add the resultant inequalities together:

$$(a_{11} y_1^0 x_1^0 + \cdots + a_{1n} y_1^0 x_n^0) + b_1 y_1^0 + (a_{21} y_2^0 x_1^0 + \cdots + a_{2n} y_2^0 x_n^0) + b_2 y_2^0 + \cdots + (a_{m1} y_m^0 x_1^0 + \cdots + a_{mn} y_m^0 x_n^0) + b_m y_m^0 \geq 0 \quad (10.4)$$

(note that since we are multiplying the inequalities by nonnegative numbers, the sense of the inequalities does not change). In exactly the same way, we multiply the first inequality of (10.3) by x_1^0, the second by x_2^0, etc., then add:

$$(a_{11}y_1^0x_1^0 + \cdots + a_{m1}y_m^0x_1^0) + c_1x_1^0 + (a_{12}y_1^0x_2^0 + \cdots + a_{m2}y_m^0x_2^0) + c_2x_2^0 \\ + \cdots + (a_{1n}y_1^0x_n^0 + \cdots + a_{mn}y_m^0x_n^0) + c_nx_n^0 \leq 0\,. \quad (10.5)$$

In both cases the sum of all terms in parentheses is the sum of all terms of the form $a_{ij}y_i^0x_j^0$ for $i = 1, 2, \ldots, m$, $j = 1, \ldots, n$. Consequently, these two expressions are equal. But since the left side of the inequality (10.5) must be less than or equal to the left side of (10.4), we have

$$c_1x_1^0 + \cdots + c_nx_n^0 \leq b_1y_1^0 + \cdots + b_my_m^0\,,$$

or $f_0 \leq g_0$, and the proof of the lemma is complete.

Step 2. *The reduction of problems A and A′ to the solution of certain systems of inequalities.*

Consider the following combined system of inequalities:

$$\begin{array}{l|l|l}
a_{11}x_1 + \cdots + a_{1n}x_n & & + b_1 \geq 0 \\
\cdots\cdots\cdots & \cdots\cdots\cdots & \cdots\cdots \\
a_{m1}x_1 + \cdots + a_{mn}x_n & & + b_m \geq 0 \\
\hline
 & a_{11}y_1 + \cdots + a_{m1}y_m & + c_1 \leq 0 \\
 & \cdots\cdots\cdots & \cdots\cdots \\
 & a_{1n}y_1 + \cdots + a_{mn}y_m & + c_n \leq 0 \\
\hline
c_1x_1 + \cdots + c_nx_n & -b_1y_1 - \cdots - b_my_m & \geq 0\,.
\end{array} \qquad (10.6)$$

As is evident from the above, this system is composed of the system (10.2), the system (10.3), and the inequality $f - g \geq 0$. The unknowns of (10.6) are the $n + m$ variables $x_1, \ldots, x_n, y_1, \ldots, y_m$. We shall prove the following assertion about this system.

If (10.6) *has the nonnegative solution* $(x_1^0, \ldots, x_n^0, y_1^0, \ldots, y_m^0)$, *then* $(x_1^0, \ldots, x_n^0)$ *is a solution to problem A*, $(y_1^0, \ldots, y_m^0)$ *is a solution to problem A′, and* $f_0 = g_0$.

Let us pause for a moment to appreciate the significance of the above statement. It is indeed remarkable that *the general problem of linear programming, that is, the maximization problem, reduces to the solution of a system of linear inequalities without any maximization requirements.* Of course, the solution of the system (10.6) (in the region of non-negative values of the unknowns) is no easier than the solution of the original problem of linear programming (problem A); however, the fact that such a reduction is possible is very interesting.

We shall now verify the above assertion. First of all it is clear that the numbers $x_1^0, \ldots, x_n^0$ are nonnegative and satisfy (10.2); similarly, the numbers $y_1^0, \ldots, y_m^0$ are nonnegative and satisfy (10.3). In addition, we have

$$f_0 \geq g_0$$

(as a result of the last inequality in (10.6)). On the other hand, step 1 yields

$$f_0 \leq g_0 .$$

Consequently, $f_0 = g_0$.

Furthermore, if $(x_1, \ldots, x_n)$ is *any* nonnegative solution of (10.2), step 1 implies that

$$f(x_1, \ldots, x_n) \leq g_0 .$$

Combining this with $f_0 = g_0$, we obtain $f(x_1, \ldots, x_n) \leq f_0$, from which it follows that f_0 is the maximum value of f.

Similarly, if $y_1, \ldots, y_m$ is any nonnegative solution of (10.3), step 1 gives

$$f_0 \leq g(y_1, \ldots, y_m) .$$

By combining this with $f_0 = g_0$, we find that $g_0 \leq g(y_1, \ldots, y_m)$; that is, that g_0 is the minimum value of g. Thus the proof of the assertion is complete.

Step 3: *Completion of the proof.* It remains to prove that if the original problem A has a solution, then (10.6) has a nonnegative solution. For then, as was shown above, $f_0 = g_0$; that is, $\max f = \min g$.

Let us assume the contrary, that (10.6) has no nonnegative solutions. If we rewrite some of the inequalities of (10.6) so that all have the sense ≥ 0, we may apply corollary 8.1 to the new system (10.7):

$$\begin{array}{l|l|l}
a_{11}x_1 + \cdots + a_{1n}x_n & & +b_1 \geq 0 \\
\cdots\cdots\cdots & \cdots\cdots\cdots & \cdots\cdots \\
a_{m1}x_1 + \cdots + a_{mn}x_n & & +b_m \geq 0 \\
\hline
 & -a_{11}y_1 - \cdots - a_{m1}y_m & -c_1 \geq 0 \\
 & \cdots\cdots\cdots & \cdots\cdots \\
 & -a_{1n}y_1 - \cdots - a_{mn}y_m & -c_n \geq 0 \\
\hline
c_1x_1 + \cdots + c_nx_n & -b_1y_1 - \cdots - b_my_m & \geq 0 .
\end{array} \qquad (10.7)$$

By assumption, (10.7) has no nonnegative solution, so by corollary 8.1 there are $m + n + 1$ nonnegative numbers $k_1, \ldots, k_m, l_1, \ldots, l_n, s$ satisfying (10.8), (10.9), and (10.10):[3]

$$\begin{array}{c} a_{11}k_1 + \cdots + a_{m1}k_m + c_1 s \leq 0 \\ \cdots\cdots\cdots\cdots\cdots\cdots \\ a_{1n}k_1 + \cdots + a_{mn}k_m + c_n s \leq 0\,, \end{array} \qquad (10.8)$$

$$\begin{array}{c} -a_{11}l_1 - \cdots - a_{1n}l_n - b_1 s \leq 0 \\ \cdots\cdots\cdots\cdots\cdots\cdots \\ -a_{m1}l_1 - \cdots - a_{mn}l_n - b_m s \leq 0\,, \end{array} \qquad (10.9)$$

$$b_1k_1 + \cdots + b_mk_m - c_1l_1 - \cdots - c_nl_n < 0\,. \qquad (10.10)$$

We shall first show that the number s is not zero. Under the assumption that $s = 0$, we consider any nonnegative solution $x_1^0, \ldots, x_n^0$ of the system (10.2) and any nonnegative solution $y_1^0, \ldots, y_m^0$ of the system (10.3). Since $(x_1^0, \ldots, x_n^0)$ is a solution of each of the inequalities in (10.2), it must also satisfy the linear combination of these inequalities formed by multiplying the first inequality by k_1, the second by k_2, and so on. Thus,

$$\begin{aligned} k_1(a_{11}x_1^0 + a_{12}x_2^0 + \cdots + a_{1n}x_n^0 + b_1) \\ + k_2(a_{21}x_1^0 + a_{22}x_2^0 + \cdots + a_{2n}x_n^0 + b_2) + \cdots \\ + k_m(a_{m1}x_1^0 + a_{m2}x_2^0 + \cdots + a_{mn}x_n^0 + b_m) \geq 0\,, \end{aligned}$$

or, combining "like" terms,

$$\begin{aligned} (a_{11}k_1 + a_{21}k_2 + \cdots + a_{m1}k_m)x_1^0 + (a_{12}k_1 + a_{22}k_2 + \cdots + a_{m2}k_m)x_2^0 + \cdots \\ + (a_{1n}k_1 + a_{2n}k_2 + \cdots + a_{mn}k_m)x_n^0 \\ + b_1k_1 + b_2k_2 + \cdots + b_mk_m \geq 0\,. \end{aligned} \qquad (10.11)$$

Since s is assumed to be zero, (10.8) states that the coefficient of each x_i^0 in the above expression (10.11) is less than or equal to zero. Because

3. $k_1, \ldots, k_m, l_1, \ldots, l_n, s$ are precisely the numbers by which we multiply the first, second, ..., $(m + n + 1)$th inequality of the system (10.7), in order to obtain (after adding) an inequality of the form

$$p_1x_1 + \cdots + p_nx_n + q_1y_1 + \cdots + q_my_m + r \geq 0\,,$$

where $p_1, \ldots, p_n, q_1, \ldots, q_m \leq 0$ and $r < 0$. The numbers appearing on the left side of the inequalities in (10.8), (10.9), and (10.10) are expressions for the coefficients $p_1, \ldots, p_n, q_1, \ldots, q_m$, and the free term r.

the numbers $x_1^0, \ldots, x_n^0$ are nonnegative, the top two lines of (10.11) sum to a nonpositive number, forcing

$$b_1k_1 + b_2k_2 + \cdots + b_mk_m \geq 0\,. \tag{10.12}$$

Similarly, the numbers $y_1^0, \ldots, y_m^0$ satisfy the linear combination of the inequalities in (10.3) formed by multiplying the first inequality by l_1, the second by l_2, and so on. Again combining "like" terms, we find that

$$\begin{aligned}(a_{11}l_1 + a_{12}l_2 + \cdots + a_{1n}l_n)y_1^0 + (a_{21}l_1 + a_{22}l_2 + \cdots + a_{2n}l_n)y_2^0 + \cdots \\ + (a_{m1}l_1 + a_{m2}l_2 + \cdots + a_{mn}l_n)y_n^0 + c_1l_1 + c_2l_2 + \cdots + c_nl_n \leq 0\,.\end{aligned} \tag{10.13}$$

Again, because s was assumed to be zero, (10.9) states that the coefficient of each y_j^0 in (10.13) is nonnegative, forcing

$$c_1l_1 + c_2l_2 + \cdots + c_nl_n \leq 0\,. \tag{10.14}$$

But the inequalities (10.12) and (10.14) contradict (10.10).

Therefore s is not zero, and it follows from (10.8) that $(k_1/s, \ldots, k_m/s)$ is a nonnegative solution to the system (10.3), from (10.9) that $(l_1/s, \ldots, l_n/s)$ is a nonnegative solution to the system (10.2), and from (10.10) that these solutions satisfy $g - f < 0$. This, however, contradicts step 1. Therefore the assumption that (10.6) has no nonnegative solution leads to a contradiction. Such a solution must therefore exist and the proof of the duality theorem is complete.

EXAMPLE. Find the maximum value of the function

$$f(x_1, x_2, x_3) = 2x_2 + 12x_3$$

under the condition that the variables x_1, x_2, x_3 are nonnegative and satisfy the inequalities

$$\begin{aligned} x_1 - x_2 - x_3 + 2 &\geq 0\,, \\ -x_1 - x_2 - 4x_3 + 1 &\geq 0\,. \end{aligned}$$

Solution. We label the stated problem A. The dual problem A' can be formulated as follows: find the minimum value of the function

$$g(y_1, y_2) = 2y_1 + y_2$$

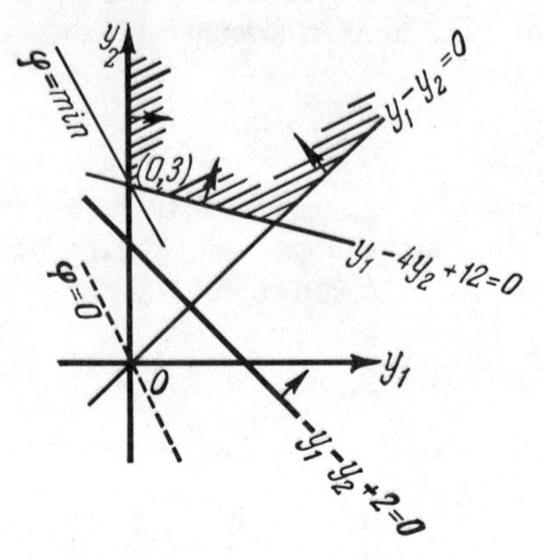

Fig. 10.2

under the conditions that the variables y_1, y_2 are nonnegative and satisfy the inequalities

$$\begin{aligned} y_1 - y_2 &\leq 0\,, \\ -y_1 - y_2 + 2 &\leq 0\,, \\ -y_1 - 4y_2 + 12 &\leq 0\,. \end{aligned} \tag{10.15}$$

Problem A' can be solved graphically by representing the solution domain of (10.15) on the $y_1 y_2$ plane; this is done in figure 10.2. From this sketch it is clear that the function g attains its minimum at the point (0, 3)—one of the vertices of the region. The value of g at this point is 3. By the duality theorem, the maximum value of the function f must also be 3.

References

1. H. Weyl, "Elementarnaya teoriya vypuklykh mnogogrannikov," in *Matrichnye igry* ["Elementary Theory of Convex Polyhedra," in *Matrix Games*]. Moscow: Fizmatgiz, 1961, pp. 254–73.

   This essay is a Russian translation of a 1935 article published in the Swiss scientific journal, *Commentarii Mathematica Helvetica*. On a very accessible level, it presents fundamentals of the theory of systems of linear inequalities, the connection between inequalities and convex polyhedra, and various theorems concerning the structure of convex polyhedra.

2. D. Gale, *The Theory of Linear Economic Models* (New York: McGraw-Hill, 1960).

   This book is a text on linear programming, the theory of "matrix games," and mathematical economics. Chapter 2, entitled "Linear Algebra," is directly concerned with linear inequalities, and the most important theorems are proved there. We remark in passing that a careful reading of the material in this chapter will enable the reader to orient himself to the more specific literature on linear inequalities.

3. A. S. Solodovnikov, *Vvedenie v lineinuyu algebru i lineinoe programmirovanie* [*Introduction to Linear Algebra and Linear Programming*]. Moscow: "Prosveshchenie" Publishing House, 1966.

   A text book on the subject under consideration, written in an accessible fashion.

4. S. N. Chernikov, *Lineinye neravenstva* [*Linear Inequalities*]. Moscow: "Nauka" Publishing House, 1968.

   A monograph on linear inequalities written by a famous specialist in this field. By means of a thorough study of the theory of linear inequalities, he acquaints the reader with various applications of this theory.